20 Makey Makey Projects for the Evil Genius

Colleen Graves

Aaron Graves

New York   Chicago   San Francisco   Athens   London   Madrid
Mexico City   Milan   New Delhi   Singapore   Sydney   Toronto

Cataloging-in-Publication Data is on file with the Library of Congress

McGraw-Hill Education books are available at special quantity discounts to use as premiums and sales promotions or for use in corporate training programs. To contact a representative, please visit the Contact Us page at www.mhprofessional.com.

20 Makey Makey Projects for the Evil Genius

2 3 4 5 6 7 8 9   LWI   22 21 20 19 18 17

ISBN   978-1-259-86046-1
MHID   1-259-86046-9

This book is printed on acid-free paper.

Sponsoring Editor Michael McCabe	**Copy Editor** James Madru
Editorial Supervisor Donna M. Martone	**Proofreader** Claire Splan
Production Supervisor Lynn M. Messina	**Indexer** Claire Splan
Acquisitions Coordinator Lauren Rogers	**Art Director, Cover** Jeff Weeks
Project Manager Patricia Wallenburg, TypeWriting	**Composition** TypeWriting

20 Makey Makey Projects
for the Evil Genius

To Val and Viv:
Thanks for being a constant inspiration
in creative problem-solving techniques!
Colleen and *Aaron*

About the Authors

Colleen Graves is a high school librarian, obsessed with learning commons transformations, makerspaces, technology education, and making stuff. Colleen was awarded the School Library Journal School Librarian of the Year Co-finalist Award in 2014, and named an Innovator by the Library Journal Mover and Shaker award in 2016 because she brings a passionate, artistic energy to the school library world. Check out her blog about makerspaces and libraries at colleengraves.org. Plus, look for her maker-focused books, coauthored with Aaron Graves, *The Big Book of Makerspace Projects: Inspiring Makers to Experiment, Create, and Learn* and *Challenge Based Learning in the School Library Makerspace.*

Aaron Graves is a school librarian with 18 years of experience in education. He is a mash-up of robot geek, book lover, and tech wizard. Aaron gained his perseverance for projects through collaborative and interactive art experiences as a member of the Good/Bad Art Collective. He is an active speaker and presenter on libraries, makerspaces, and research skills. In his free time he can be found writing, restoring microcars, or inventing something that makes you smile!

Contents at a Glance

SECTION FOUR   About Makey Makey Go

Contents

SECTION ONE   Fun and Games

SECTION TWO   Interactive

Foreword

INSTEAD OF A TRADITIONAL FOREWORD, Jay Silver and Eric Rosenbaum thought it would be fun to have a casual chat and enlighten readers about the origins of Makey Makey and discuss the big ideas behind this little invention kit.

COLLEEN AND AARON: *How did you come up with Makey Makey?*

JAY: We were super obsessed with the everyday world, skin to skin contact, and touching things in nature.

ERIC: During that time I was making musical instruments out of fruit. And generally obsessed with inventing crazy new musical experiences (and improvisational experiences) for people to play with. We were both on the Scratch team helping kids bring their creations to life. Also, we were hacking keyboards for one of our music projects. This got us thinking about Scratch and how it's incredibly flexible in how it lets you manipulate stuff on the screen: images, sounds, animations, stories. Scratch has this powerful flexibility and open endedness. We also wanted that kind of flexibility for physical stuff and how to hook things up in a super flexible way.

JAY: Yeah, like digital duct tape. Or physical digital duct tape that would bridge the real world with the digital world. What could we make that would zip tie apps to junk? And then how do you think of junk as computational? Computation is the greatest practical power but it is tied up in coding. And coding has barriers

to access no matter how simple you make it. What if you could cut an app in half and cut junk in half and splice it together?

ERIC: We also thought a lot about making things tinkerable. One way to do that is to make them really immediate, so you could try something and see its effect right away. We had this experience of working with electronics and Arduino but there were a lot of steps in between. So the idea of the keyboard hack and making a device that thinks it's a keyboard partly came from that. Because you just plug it in and try something and before you know it your computer makes a fart sound because it's so easy.

Being MIT students, we could hack these things together, but even for us having a radically easier, faster, simpler way to do it was something we really wanted.

JAY: In Lifelong Kindergarten (LLK) we're obsessed with creating the building blocks and fingerpaints for people of all ages.

The reason we were in LLK was because we want to live in a world created by everybody. That can only be possible if you create tools that help people create the world (and those tools are very easy to use.) Not only does everyone then get to create the world, but then you get this added benefit that once you get used to the tool, then you become virtuosic with it.

Anyone can walk up to a piano and make some sounds, but if you practice like thousands of hours then you can make stuff up on the fly and it will sound like it was composed. We wanted to make these kinds of "instruments" that anyone could walk up to and make inventions with. Anyone hook up a banana piano in minutes.

COLLEEN AND AARON: *How did you figure out how to make Makey Makey?*

ERIC: A lot of tinkering went into it. The first ones were built in collaboration with The Tinkering Studio. We made some prototypes by hand and tried them out there.

JAY: Eric and I were messing with Drawdio in the lab and one day we sat down and drew out all of the physical possibilities of how we could hook up Drawdio with icons and then we videoed each icon. Then we thought, "How could we hook this up to a computer?"

We started messing around with keyboard hacking where you take apart a regular keyboard and hook it to a homemade button.

Even before that, Eric was a part of the Playful Invention Network, which was an NSF funded project connecting science museums to the Lifelong Kindergarten group and they were making fruit and ice instruments of different stuff.

ERIC: That was in 2001 when I was doing that.

JAY: So all of these streams merged together. And Makey Makey is only one concrete way of getting these ideas out to people. You have to make something that's easy enough to use, to make, affordable to buy, etc.

So the summer of 2010 all these things came together and Eric made a prototype. We thought we could Kickstart Makey Makey and launch it at Maker Faire.

ERIC: When we talk about connecting the everyday world to Makey Makey, one category we're talking about is conductive stuff: almost anything that you can eat, or that is alive, or used to be. And things that are wet or metallic.

JAY: A lot of things in the everyday world don't conduct electricity at all and it takes both of those to make up a Makey Makey project.

ERIC: Right, if you didn't have anything non-conductive nothing would work!

JAY: If the air was conductive, then you could never hook up a Makey Makey because the key press would never stop! Hahahaha

COLLEEN AND AARON: *Why does the original Makey Makey complete a circuit instead of using capacitive sensing?*

ERIC: One crucial thing is that capacitive sensing requires calibration to get into the right range. This ties directly into our idea with the Makey Makey of wanting extreme ease of use and flexibility. We wanted it to work all the time no matter what you throw at it. The resistance sensing style that we created for Makey Makey is much more stable. Thanks to Jay's digital filtering technique that stabilizes the whole thing. He did MatLab simulations of the noise and signal and then I implemented the digital filter on the Arduino.

JAY: The beautiful thing about that was that I was trying to test out the DSP filters at Yogaville and I was trying to find the noisiest environment I could. Electrically noisy environments are best found in big cities with big appliances or industrial equipment and there was almost nothing electrically noisy in Yogaville.

ERIC: But I was on the phone with him here and it was super electrically noisy here at Media Lab so we tested Makey Makey against noise here in Cambridge, Massachusetts.

JAY: We had to run an Internet cable 500 feet through the forest to get an Internet connection in Yogaville. We had a splice box in a tree inside some ziploc bags and a tupperware container.

ERIC: I remember, I didn't understand how the filtering was actually going to remove the noise, when you were trying to describe it to me. And then you showed me a graph and I was amazed.

COLLEEN AND AARON: *What are some things you cut from the Kickstarter video?*

ERIC: We found a gong and we wanted to be able to dislike a song on Pandora by hitting the gong to click the thumbs down on Pandora. We tried to film it, but we just couldn't get it to read right. It wasn't clear what was happening. Also, I guess most people don't have gongs.

JAY: It was very important to us that in every single scene the viewer could visually digest and imagine herself building the project. It was also important to us that every scene left some room for the person watching the video to enjoy discovering some of their own ideas. So that after watching it, people would think, "Oh wait . . . I have an idea I want to try."

ERIC: Right, we wanted people to connect with the ideas in the video and realize, "Oh I have those things in my house "

JAY: Even for people who never buy Makey Makey, we want people to think, "They just showed me 8 or 9 things that could be made and I have ideas in between those things and outside of the box that those things are in and I think I wanna go make something." That's why we used stairs, bananas, and buckets of water. Actually, those weren't buckets, they were plastic drawers from a cheap storage unit. We actually had a scene that we cut where those drawers were full of clothes and we threw the clothes out before filling them up with water.

ERIC: Because we actually had to do that, right? Those drawers were full of craft materials or something.

JAY: Yes, so every scene was shot with the idea that we should choose something that's iconic so people understand it visually but also feel like they have access to this stuff too. And that inventing is really as easy as just alligator clipping two things together and that everyone's an inventor. Our only real message with making the product and the way the video is shot is to let the world know that everyone is an inventor. Maybe not for a living. But an inventor nonetheless.

ERIC: During the video shoot, it took a lot of takes to get the cat to behave because he didn't want to sit on the tinfoil. In addition, somehow through a statistical fluke our can of alphabet soup had no y's and we had to make them out of x's. I had to cut one leg off of the x's to make y's to spell Makey Makey out of alphabet soup noodles!

JAY: Eric and I wrote the scenes over the phone. Eric flew in and we set up two cameras, two clamp lights, diffusers made out of paper towels, and we shot the entire video in our house or our yard. So the projects would feel doable (but it was also the limit of our resources). All of the sounds came from our mouths. I made the sound effects of the alligator clips clipping onto things with my mouth and Eric sung the intro jingle. Everything we had in the video came out of our house. We edited, videoed, directed, etc. the whole thing ourselves.

JAY ASKS COLLEEN AND AARON: *Of all products you could have written about, why did you write a Makey Makey book?*

COLLEEN: I became obsessed with the idea of the Makey Makey after watching a bananaphone video because I thought I'd be able to hack my daughter's toy and make it sound heavy metal (that's actually a project in this book!). I'd been tinkering around with Arduino with little success and so I started learning more about coding by going through the code. org lessons and learning to use Scratch and I finally got to play and borrow a few Makey Makeys from my librarian friend, Leah Mann. I brought them to my middle school at the time and started getting really wacky playing with Makey Makey and thus entering the gateway to physical computing. The next summer Dave asked me to write lesson plans for Makey Makey and at that point, I started pushing myself to try further possibilities and get wicked inventive with this awesome little controller. It gave me creative confidence in making with electronics and changed my worldview. Now I feel like I can actually start hacking things. I look at everyday stuff and think, "How can I make my own version of that with Makey Makey?"

AARON: We also believe that our goals as educators align with your goals as inventors. We think making and invention should be accessible for all people. We appreciate the purposeful approachability that you built into Makey Makey. Plus, it's easy to quickly make something with a Makey Makey, but you can get more complicated quickly if you want to as well. We love that it has a low floor, wide walls, and high ceiling with loads of possibilities.

Lastly, it engages students and teachers across the curriculum and allows educators to let students guide their own learning by learning with their hands instead of listening to a Science lecture. There is no other tool (other than pencil and paper) that I can give to a teacher of any subject or any grade level and they can quickly find success with it. That's a rare thing and one of the reasons we love Makey Makey!

COLLEEN: Plus it's the gateway to making. Makey Makey pushes you to want to learn new things. How can I write a program that does this? How can I build more switches out of different materials? What kind of thing can I make that needs a switch? Or could easily be interfaced with Scratch?

JAY: *Aaron, was the pinball machine really hard to make?*

AARON: There were like four versions of it, so it should be more easily made now. We had this idea that the ball would be the trigger for stuff, but the ball was moving so fast that it really wasn't registering. Then we found it would work if we slowed it down enough. But that's the trick—can you slow a ball down going down the rails long enough to make it make a key press? Plus, it was supposed to be out of cardboard, but I didn't read Colleen's pitch for the project and I just started making it out of wood. I was trying to design something that was simple enough for anyone to make even if they had a limited amount of tools. As a design constraint, this caused me to build the pinball machine over and over and over again.

COLLEEN: The other one you did wrong was the cookie jar. That one was supposed to be a lock that you unlocked by exercising, but instead you just made an alarm. Which was actually better though because we needed more simple projects. Plus, that allowed us to make a challenge of hacking the cookie jar project with the lock box project.

We also wanted to respect the idea that everyone is an inventor. So our projects are just ideas of how to do the projects one way, but if you want to try and do them another way, then go do that!

JAY: *What kind of creative response would you be excited to see when the book comes out?*

AARON: We like to see people sort of following our instructions, but hacking them on their own too. It's cool for me to see a project and think, "Oh, that's something I started, but someone has taken it and done something totally different or modified it." That's the coolest thing.

Acknowledgments

Thanks to Jay Silver and Eric Rosenbaum for making this awesome digital duct tape that allows anyone to become an inventor! We are so thankful to you for taking the time to write an interesting and informative foreword to this book. A bucket full of thank yous, goes out to Jay for always being just a phone call away to chat about high-level Makey Makey stuff.

Special thanks to Liam Nilsen for giving us feedback on our book projects and for responding to that first tweet that started our Makey Makey collaborations!

Thanks to Dave Ten Have for believing in me early on and enlisting my help for the first Makey Makey Lesson plans. (Colleen)

Jie Qi, we appreciate the time you spent discussing Colleen's quirky ideas at mashing up Chibitronics and Makey Makey.

We are forever grateful to Susan and Sterling Price for giving us time to work on our creative ideas and being great grandparents to our little peeps!

Once again thanks to Josh Burker for being awesome, pushing our thinking, and helping us problem solve when we got stuck during our projects!

Kristi Taylor, thank you so much for your help with Illustrator files. We appreciate you more than you know!

This book wouldn't be possible if it weren't for Michael McCabe, our editors at McGraw-Hill, and the ever-amazing Patty Wallenburg! Thank you for all you do that made our dream book into a reality.

And lastly, to Denton's Shift Coffee, thanks for providing amazing lavender lattes and a comfortable work environment so we could tune in and focus on writing.

An Introduction to Physical Computing

PHYSICAL COMPUTING IS THE IDEA of combining actions in the real world with actions on a computer. Since the Makey Makey Invention Kit relies on physical actions and everyday objects to trigger a response via your computer, we consider it the gateway to physical computing. The projects in this book should get your feet wet in physical computing, fill up your coding toolbox, and get you on your way to even more computational tinkering with Raspberry Pi and Arduino microcontrollers.

Banana Mash-Up

This book also focuses on mashing up all kinds of other cool stuff with your Makey Makey to really get you the most for your banana. These projects will give you lots of different ideas of things to try to use your Makey Makey with the outside world. You'll mix up Makey Makey with 3D printing, paper circuits, Scratch programming, Processing, and even turning everyday trash and recycling materials into new and cool electronic toys. We hope that by connecting these computational ideas and building up your skills and tool kit, you might find a new passion. Maybe you'll see how easy it is to write code in Processing and decide to delve further into learning this programming language. Or maybe you'll find a new passion in creating wearable electronics. Who knows? Just get started making and see where your learning takes you!

Some of these projects focus on heavy building but with simple computing (the pinball machine is one such example), while others will be simple builds but really take your Scratch programming skills to the next level. (Try the swipe input project to flex your programming skills.) To help you decide where to start, we've labeled each project by cost, time, and skill level.

Project Layout

Every project includes this info:

Cost

Make time

Skill level

Supplies

Cost Explained: Cost Beyond Makey Makey

$	Under $20
$$	$20–50
$$$	$50–75
$$$$	$75–125

Skill Level Explained: One Banana, Four Bananas

Skill Level Ratings for Difficulty

🍌	A beginner project that focuses on building creative confidence. A plug-and-play project requiring the basic Makey Makey kit with minimal supplies. Simple programming or a basic program remix.
🍌 🍌	These projects require delving into some light physical computing with intermediate builds and light levels of programming.
🍌 🍌 🍌	Some projects may require advanced skills such as soldering but still have intermediate to light programming.
🍌 🍌 🍌 🍌	A complex build or program that requires advanced skills or advanced tools such as a sewing machine, saw, or other fabrication tools.

One-Banana Hack

When applicable, we include this quick and easy way to make a four-banana project into a one-banana project.

Word on Safety: Makey Makey is very safe product and complies with Federal Communications Commission (FCC) regulations. Initially, some users may worry about being shocked, but the amount of electricity being used is very low, and shocks will not happen. We urge you to wear safety goggles while working on more complex builds that use tools and to follow safety tips and practices listed for some of the individual projects. Most of all, just as the warning on the box states: "Users may start to believe they can change the way the world works. Extended usage may result in creative confidence."

List of Projects

Section 1

- Project 1: Makey Makey Spin Art with Motors
- Project 2: Cootie-Catcher Paper Circuit with Makey Makey
- Project 3: Makey Makey Marble Maze
- Project 4: Makey Makey Arcade Coin Slot
- Project 5: Arcade-Style Fortune Teller
- Project 6: Build Your Own Pinball Machine

Section 2

- Project 7: Cookie Jar Alarm
- Project 8: Makey Makey Light-Up Morse Code Tower
- Project 9: Makey Makey Etch-a-Sketch with Processing
- Project 10: Makey Makey Musical Hoodie
- One-Banana Hack of Musical Hoodie: Hack a Plushie
- Project 11: Makey Makey Swipe Input

Section 3

- Project 12: Hacking a Kid's Toy with Makey Makey
- Project 13: One-Banana Hack: Hacking a Kid's Toy
- Project 14: Makey Makey Power Tail Prank
- Project 15: Makey Makey Lock Box

Section 4

- Project 16: Makey Go No Donut Prank
- Project 17: Makey Go "Chopsticks"
- Project 18: Makey Go "Heart and Soul" Plant Kalimba

- Project 19: Makey Go Cat Clicking Game
- Project 20: Makey Go Lemon Squeezy

List of Projects by Skill Level

One-Banana Projects

- (Section 1) Project 1: Makey Makey Spin Art with Motors
- (Section 2) Project 7: Cookie Jar Alarm
- (Section 2) One-Banana Hack of Project 10
- (Section 2) One-Banana Hack of Musical Hoodie: Hack a Plushie
- (Section 3) Project 13: One-Banana Hack: Hacking a Kid's Toy
- (Section 4) Project 16: Makey Go No Donut Prank

Two-Banana Projects

- (Section 1) Project 3: Makey Makey Marble Maze
- (Section 1) Project 4: Makey Makey Coin Slot Scratch Hack
- (Scction 2) Project 8: Makey Makey Light-Up Morse Code Tower
- (Section 2) Project 9: Makey Makey Etch-a-Sketch with Processing
- (Section 3) Project 14: Makey Makey Power Tail Prank
- (Section 4) Project 17: Makey Go "Chopsticks"
- (Section 4) Project 19: Makey Go Cat Clicking Game

Three-Banana Projects

- (Section 1) Project 2: Cootie Catcher Paper Circuit (requires soldering)
- (Section 1) Project 5: Arcade-Style Fortune Teller
- (Section 2) Project 11: Makey Makey Swipe Input
- (Section 4) Project 18: Makey Go "Heart and Soul" Plant Kalimba
- (Section 4) Project 20: Makey Go Lemon Squeezy

Four-Banana Projects

- (Section 1) Project 6: Build Your Own Pinball Machine
- (Section 2) Project 10: Makey Makey Musical Hoodie
- (Section 3) Project 12: Hacking a Kid's Toy with Makey Makey
- (Section 3) Project 15: Makey Makey Lock Box

20 Makey Makey Projects
for the Evil Genius

SECTION ONE
Fun and Games

A true evil genius values his or her own entertainment above all others, and that is the objective of the projects in this section. Whether it's passing the time by making spin art or challenging minions on a homemade pinball machine, the projects in this section are as fun to create and modify as they are to play!

- **Project 1:** Makey Makey Spin Art with Motors
- **Project 2:** Cootie-Catcher Paper Circuit with Makey Makey
- **Project 3:** Makey Makey Marble Maze
- **Project 4:** Makey Makey Arcade Coin Slot
- **Project 5:** Arcade-Style Fortune Teller
- **Project 6:** Build Your Own Pinball Machine

Makey Makey Spin Art with Motors

SOMETIMES WORLD DOMINATION is thwarted by computer updates. As an evil genius, you can't let this downtime get away from you. For this project, you'll create something similar to traditional spin art, but instead you'll hack spin art with Makey Makey and sticky notes. You'll be able to trigger the spinning motor by stepping on or pressing a switch. This project requires no programming, so while your computer is down for updates, you can use a portable USB power charger or phone charger to make art no matter where you are (see Figure 1-1)!

Cost: $

Make time: 30 minutes

Skill level: 🍌

Figure 1-1   Spin-art project.

Supplies

Materials	Description	Source
CD ROM drive motor	Uxcell DC 3-V, 24-mm base car VCD DVD player spindle motor with tray holder	Recycling or Amazon
Takeout container and lid	Plastic container with lid if you don't have a CD available	Recycling
Hot-glue gun and glue sticks	Low-temperature glue gun	Craft store
CD	Damaged CD or one you're tired of listening to	Recycling
Portable USB phone charger	Optional if you want to take this show on the road and not your PC	Amazon

Step 1: Motor Size and Voltage

The Makey Makey will power a small motor using the KEY OUT pin located on the back of the controller. The best motor we found for this project is a motor from an old CD/DVD drive. (see Figure 1-2) The low-voltage operating range of 1.5 to 5 V for these motors is perfect. They were designed especially to spin a flat disk, and many of them have a mount for a CD to clip into. If you don't have an old CD/DVD drive lying around to cannibalize, these motors can be bought from electronics vendors and on Amazon.

Step 2: Motor Test

On the back of the Makey Makey you will find a few different pins in the middle header that will power a light-emitting diode (LED) or motor. First in the row is KEY OUT, which will power a motor when any key input is pressed. MS OUT will power a motor when the mouse

Figure 1-2   Scavenge for motors.

keys are signaled. Then in the middle of this header is a 5-V pin that will produce a constant 5-V output that will be helpful for testing your motors. Next to this pin is the GND pin, which will ground a LED or act as an extra ground if you need it. Lastly, PGD and PGC are used for programming at the factory, so you'll want to leave those alone. You don't want to always power your motor, so for this project you are going to use the KEY OUT output pin, which sends out a 5-V signal only when a key is pressed. You'll have to wire up a key too, but first let's determine if the Makey Makey will power the motor by connecting one of the contacts on the motor to the KEY OUT pin. Using a white jumper wire from your kit, place one end firmly into the KEY OUT pin as in Figure 1-3, and use an alligator clip to connect your jumper wire to the red wire of your motor. You will also need to ground your motor, so use another jumper wire in the GND pin. Connect

the jumper wire to an alligator clip and then the opposite end of the alligator clip to the black wire on your motor. Just to note, you don't have to use the GND pin on the back of the Makey Makey; you can actually connect your alligator clip to any EARTH input on the bottom of the Makey Makey.

When you wire your project in this way, the motor will not get a signal until a key is pressed. Test this with your hands by holding the EARTH connection on the Makey Makey and then pressing and holding any arrow key to see if the Makey Makey will provide enough output to turn the motor (see Figure 1-4). For some motors we tested, we noticed that it would be enough power to turn the motor, but once we added a sticky note or tape on top as a canvas for our art, the output wouldn't be enough to get the motor started. If you are not using a CD drive motor for testing purposes, add a little weight by using a bit of double-stick tape to adhere a couple of sticky notes to the motor shaft to test whether the motor will still turn when a key is pressed. If you determine that your motor will turn when a key is pressed, set the Makey Makey and motor aside, and leave the jumper wires in position. It's time to make a stand for your spin art.

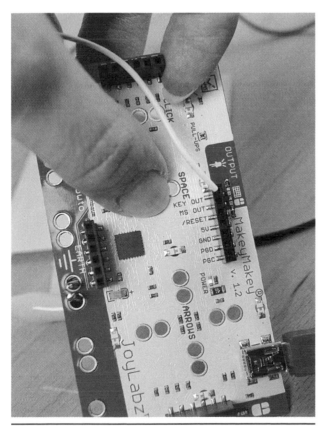

Figure 1-3 Output pins on the Makey Makey.

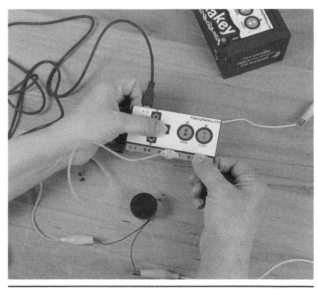

Figure 1-4 Testing the motor.

Step 3: Cold Cuts, Hot Glue, and Takeout

For the base, you will need a large plastic takeout container (think soup) or large yogurt container with a flat bottom. Turn the container over so that the base is on top, and place the motor in the center. Use a little dab of hot glue to adhere your motor to the container, as in Figure 1-5. Make a mark off to the side of the motor, and use a craft knife to cut a small V or triangle, as shown in Figure 1-6. Flip the container over, and on the bottom edge make two cuts 1 inch apart and 1 inch long. Fold the flap back, and trim it off with a pair of scissors or a craft knife, as shown in Figure 1-6.

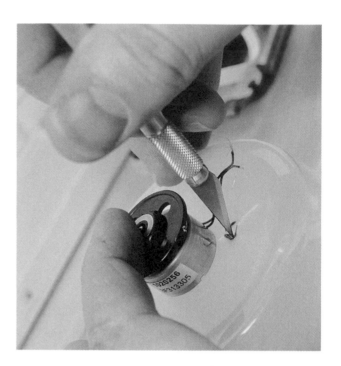

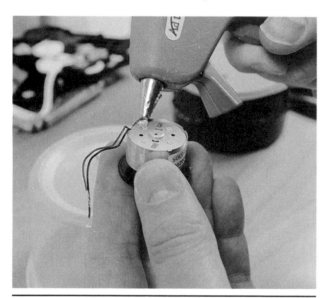

Figure 1-5 Glue the motor in place.

Flip the container back over, and feed the wires coming from the motor through the hole you just made. Inside the container, connect red and black alligator clips to the respective wires from the motor, as shown in Figure 1-7. (Color doesn't actually matter; we are just using color continuity to aide in instruction.) Connect the red alligator clip to the jumper wire in the KEY OUT pin and the black alligator clip to the jumper wire in the GND pin (or to an EARTH input on the front of the Makey Makey).

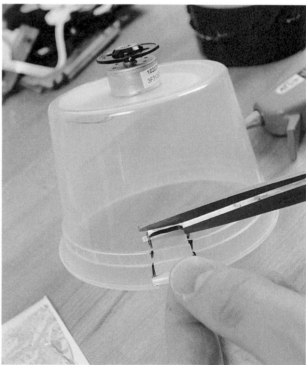

Figure 1-6 Convert the container.

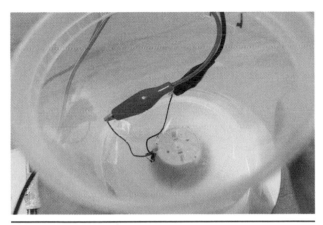

Figure 1-7   Wire the motor.

If your motor already has a mount for a CD, place a CD into the mount. If the CD does not stay in place, use a little hot glue to hold it in position, but be careful not to let the glue drip down and keep the motor from spinning (see Figure 1-8). If you are using a motor with a small pulley mount or no mount at all, use a thumbtack or nail to create a hole in a pencil eraser, and slide the eraser onto the motor shaft. Use a small amount of hot glue to carefully secure it to the shaft. Instead of using a CD, you could also hot glue the lid from the container to the motor shaft.

Step 4: Switch

You will need to create a switch so that you can activate the motor from the KEY OUT pin to make art. You can make a pressure switch by using three sticky notes of the same size. Cut and adhere a piece of foil to the bottom sticky note using double-stick tape (see Figure 1-9). Be sure to create a small foil lead on one side so that you can attach an alligator clip to it. Repeat the process for the top sticky note, and place the foil lead on the same side as you did with the bottom because when you flip the paper over, the lead will end up on the opposite side. For the insulator, use two sticky notes. Fold them over, and cut a small hole (or a few holes) that will allow the foil on the top and bottom sticky notes

Figure 1-9   Sticky note switch with foil leads.

Figure 1-8   Glue the CD if necessary.

to make contact only when the switch is pressed (see Figure 1-9). Cover the bottom side with an insulating middle paper, and make sure that no foil is exposed on the outside other than the foil lead that you created, as in Figure 1-9. Cover the top side of your switch with your other middle insulator to make sure that no foil is exposed

Figure 1-10 Connect Makey Makey and phone charger for power.

other than the lead you created here as well. (Otherwise your switch might short circuit!) Tape all three pieces together, making sure not to cover the foil with tape where the holes for your switch are located! Connect an alligator clip from the bottom foil lead to an EARTH input on the front of the Makey Makey. Attach another alligator clip to the top foil lead on your sticky note switch, and connect it to any arrow input on the front of the Makey Makey (see Figure 1-10).

Step 5: Power Up the Makey Makey

Sometimes you want to use Makey Makey without a computer. If you are doing a strictly output project like this, you can power the Makey Makey with a USB phone charger. Just plug the USB into the phone charger as in Figure 1-11.

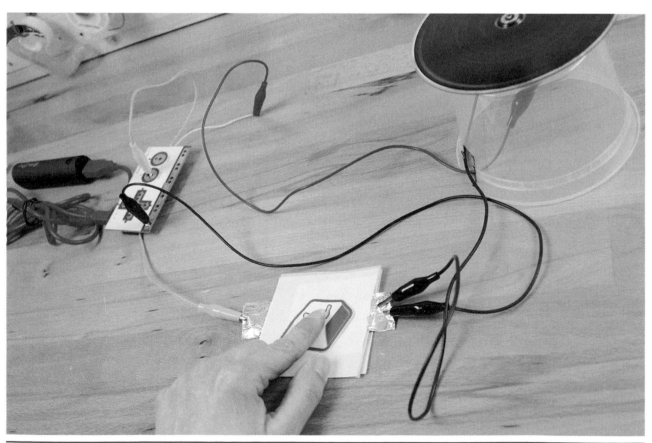

Figure 1-11 Full set up for Makey Makey spin art.

Step 6: Sticky Note Spin-Out Art

Now that you've made a switch and powered up your Makey Makey, you are ready to make some sticky note art! Center a sticky note on the CD, place your elbow on the switch, and let the motor turn until it has reached full speed. Lightly touch the sticky note with the tip of your favorite color marker, and move it slowly outward from the center to create some amazing spin art! Change colors and keep creating art!

Taking It Further

What other ways could you get inventive with switches for your output? Could you find a way to make your spin art spin by swirling water? Or make a Play-Doh switch? What about wiring up a button? The LED Morse code project (in Section 2) is another fun project that plays with this concept. You can also get some pretty wild prank ideas with the power tail switch featured in Section 3.

PROJECT 2

Cootie-Catcher Paper Circuit with Makey Makey

AS AN EVIL GENIUS, one of the hardest things to do is entertain your league of minions. Those minions get bored, quite a lot actually. Keep them entertained for hours on end by making some paper circuit cootie catchers that they can reprogram over and over again with Scratch. In this way, the minions can spend hours adding annoying noises and changing each other's fortunes on a whim (see Figure 1-12).

Cost: $

Make time: 30 minutes

Skill level: 🍌🍌🍌

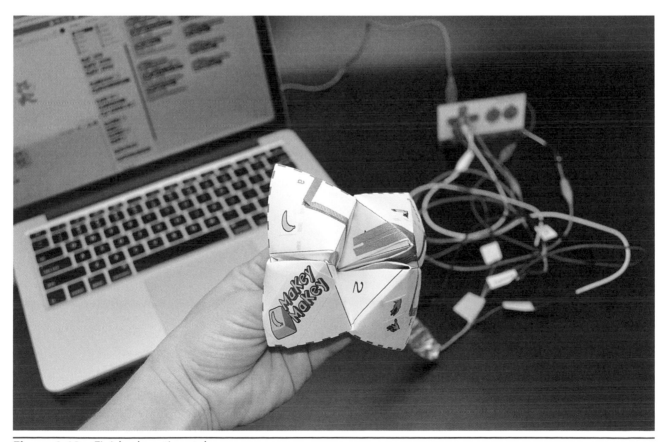

Figure 1-12 Finished cootie catcher.

Supplies

Materials	Description	Source
Office supplies	8½- by 11-inch paper, ½- to ¾-inch-wide clear tape	Office supply or craft store
Conductive tape	Fabric tape from Makey Makey Inventor Kit	Joylabz
Hookup wire	Elenco 6 Color Hookup No. 22 AWG Solid Wire Kit	Elenco
Thin-gauge hookup wire	50 feet of 30 AWG wrapping wire	Radio Shack item no. 2780502
Computer and access to Scratch	A free visual programming language and online community	scratch.mit.edu

Create Scratch Game

Step 1: Create Account and Get Started

Scratch is one of the easiest and quickest ways to start making interactive projects with Makey Makey. It's almost as if the two were made for each other! If you are new to Scratch, this project will be a good introduction, but feel free to play around and see how complicated you can get with Scratch programming. Once you've created an account, we are going to make six quick programs that will control your paper circuit cootie catcher. Create an account at https://scratch.mit.edu/. Login and click "Create" to get started on your first project. Name it "Cootie Catcher" or something similar so that you can find it again in the "My Stuff" tab. Everything you do in Scratch can be controlled by the background or by the sprites (aka characters) in your game. For this project, we are just using Scratch as a way to make sound effects for your paper circuit. However, if you want to add more sprites and tinker with changing the backdrop to create personalized animations, look at Project 5. For now, you need to know that the area on the left is the "Stage" where the action happens, and all the controlling blocks (or *scripts*) are located in the center in brightly colored menus. On the right you have your work area where you drag blocks and lock them together to create simple and complex programs.

Step 2: Program Arrow Keys in Scratch

To make sound effects activated by Makey Makey, we'll program each arrow key with funny fortunes. In the "Blocks" tab, click on the brown "Events" menu. These are the blocks you'll need to program key controls. Since Makey Makey works like an external keyboard, you'll be here often getting blocks to control your Makey Makey creations. It also houses the "When flag clicked" block that you will need at the beginning of any scripts that you want to run when the game starts. To start, drag a "When space clicked" block from the "Events" palette to your work area. Using the dropdown menu, choose the up arrow key. Navigate to the pink "Sound" menu, and drag a "Play sound meow" block and connect it to your "When up arrow key clicked" block. You are now ready to create personal sound effects in Scratch! Above the brightly colored "Blocks" menus, you might have noticed the tabs "Scripts," "Costumes," and "Sounds." Click on the "Sounds" tab to begin recording your own sounds. The microphone icon (as seen in Figure 1-13a) will allow you to record fortunes. Create four fortune recordings, and label them so that you can program your arrows accordingly. Once you have four recordings, click back to the "Scripts" tab, and

Figure 1-13   Program sounds in Scratch.

you will see them available in the "Play sound" block. Choose one sound for each arrow key, as we did in Figure 1-13b. These sounds will be featured as the fortunes in the center of your cootie-catcher paper circuit.

Step 3: Program "W" and "A"

One of the most fun aspects of creating your own fortune teller is having minions pick numbers and gamble with fortunes. We thought it would be fun to add a sound effect every time you move your cootie catcher. Go through the "Sound" menu in Scratch to find a sound effect that you want to play as you open and close your cootie catcher. You can program this the same way you did your arrow keys, but we are going to add a little trick to start our cootie-catcher game. By using the "if/else" conditional statement, we can create two conditions for one key press by adding this simple combination code! Drag a "When space key pressed" event to your work area, and using the dropdown menu, change "space" to the "w" key. Now you'll add an "if/else" statement that will allow Makey Makey to activate a different sound if both the "w" and the "a" keys are pressed at the same time, which will happen when you have your cootie catcher closed. Find the "Key space pressed?" block in the "Sensing" menu, and insert it into the hexagon-shaped space between "if" and "then." Remember to change the dropdown to "a," as shown in Figure 1-14. (As you drag the block, you will see the space highlighted, and the block will drop into place.) This program uses AND logic and will tell our Scratch game to play a different sound only when the "w" and the "a" keys are pressed at the same time. Create the sound you want to start your game, and drag the "Play sound" block inside the "if" block. If you have a problem with the sound repeating too soon, you can add a "Wait 1 secs" block from the conditional statements to allow the full length of your recording to play before the program moves to the next script. Adjust your wait times as necessary. When only the "w" key is pressed, you can have a different sound playing by dragging a "Play sound" block inside the "else" block. In this way, one sound will play if both buttons are pressed, but if only one button is pressed, the other sound will play. In order for this to work for both keys, you need to duplicate these scripts by right-clicking on the script "When 'w' key pressed" block. Make sure

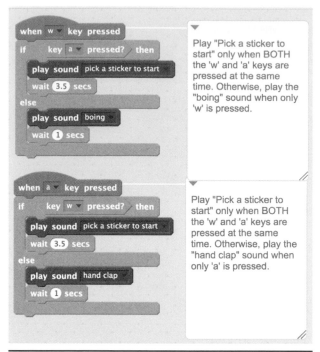

Figure 1-14   If two keys are pressed.

that you change your scripts to match Figure 1-14. Now you are ready to build your Makey Makey cootie catcher!

Create Cootie Catcher

Step 1: Print Template, Cut, and Fold

Print the template (see Figure 1-15), cut around the dotted lines, and with the backside facing up, fold the four corners into the center. Flip the paper over so that you can see the four arrows, and fold the next set of four corners into the center (see Figure 1-16). Fold the paper in half, and then unfold and fold the paper in half in the other direction. Now you can push the corners to the center and place your thumb and pointer finger into the cootie catcher. Practice opening and closing the "mouth" of the paper. Are you ready to make it interactive?

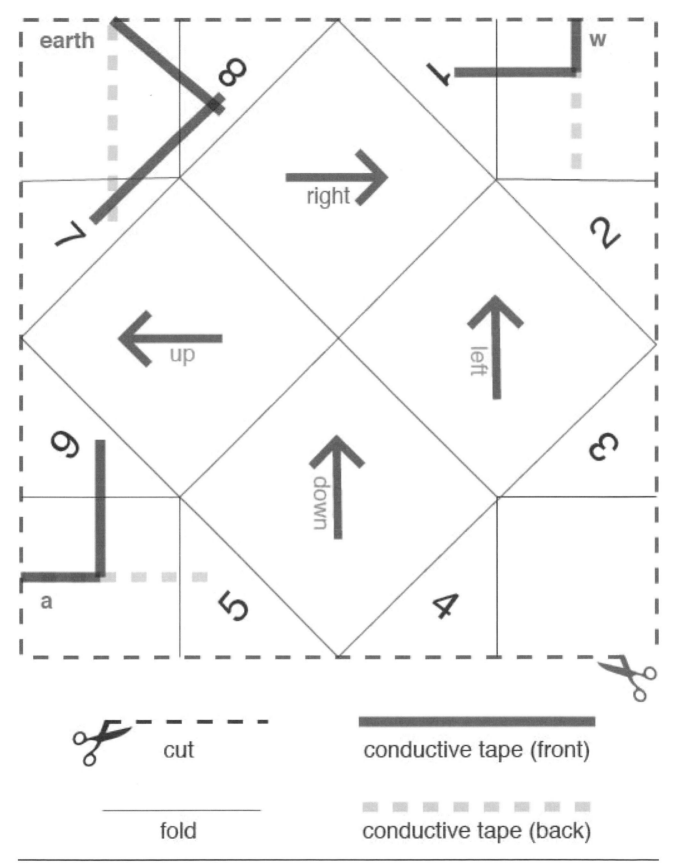

Figure 1-15 Cootie-catcher template.

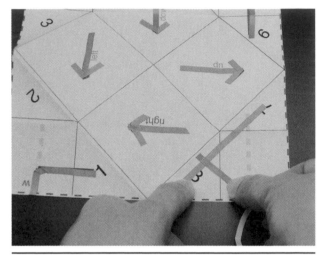

Figure 1-17 Apply conductive tape to the front.

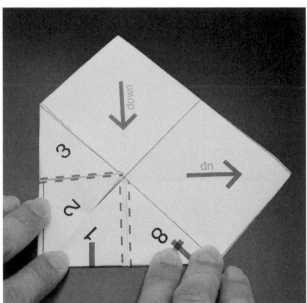

Figure 1- 16 Fold the template.

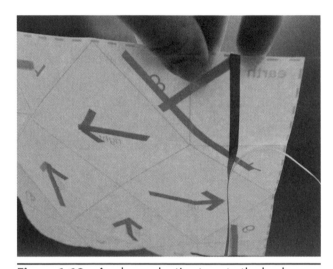

Figure 1-18 Apply conductive tape to the back.

Step 2: Conductive Tape on Arrows

To prepare your cootie catcher for Makey Makey, we have to create some conductive conditions. Open your cootie catcher to the arrows, and press it flat. Using conductive fabric tape from the Makey Makey Inventor Booster Kit, apply tape to the arrows as directed in the template and shown in Figure 1-17. Flip the paper over, and apply conductive tape to the back of the template, as in Figure 1-18.

Step 3: Refold

Refold the cootie catcher as before. Your cootie catcher should look like Figure 1-19 from the back and Figure 1-20 from the front. Once you attach the Makey Makey, the tape on the "w" will complete a circuit and create the sound effects you programmed in the first part of this project.

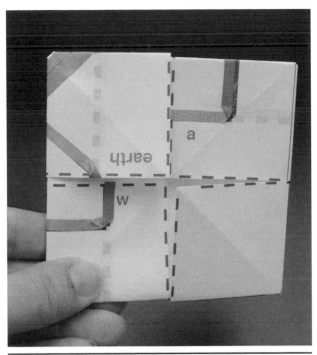

Figure 1-19 Refolded back.

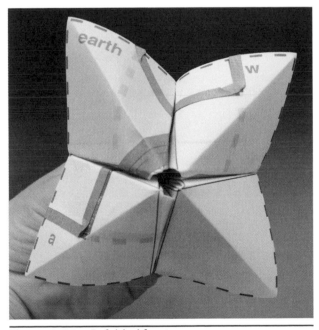

Figure 1-20 Refolded front.

Step 4: Conductive Tape on EARTH

To make sure that you get a really good connection, layer more tape over the EARTH portion of your cootie catcher. You can also use a small triangle of aluminum foil for this step, as long as you make sure that the foil is touching the EARTH tape and you do not cover the EARTH conductive tape trace with clear tape and accidentally insulate your circuit (see Figure 1-21). (The term *trace* comes from printed circuit boards [PCBs] that incorporate conductive traces instead of wiring. It is often used when creating paper or sewing circuits because a conductive trace is also used instead of wires.)

Figure 1- 21 EARTH connections.

Step 5: Prepare Wires

Because we don't want the bulkiness of the alligator clips and they just aren't long enough, we are going to create some hybrid wires for connecting our Makey Makey to this paper circuit. Cut six pieces of 16-inch-long wire. Strip both ends of the wire about 1 inch. To make hookup easier, use masking tape to clearly label what each wire will be hooked to, as shown in Figure 1-22.

Figure 1-22 Clearly labeled hookup wires.

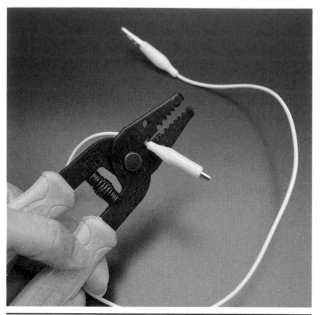

Figure 1-23 Cut the alligator clips.

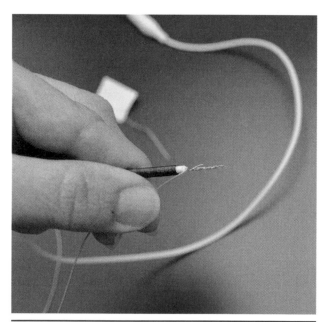

Figure 1-24 Twist the exposed wires together.

Cut the ends off of six alligator clips from your kit (see Figure 1-23), and strip the end you cut so that you have about 1 inch of copper wire exposed. Place a small piece of heat shrink on the cut end of the alligator clip, and then twist the cut end of the alligator clip to one end of the hookup wire, as in Figure 1-24. Solder the two wires together, and then cover with heat shrink, as shown in Figure 1-25. When you are done, you will have a wire that you can connect to each key press on your cootie catcher but still have an alligator clip to clip easily to the Makey Makey. In this way, when your Makey Makey is not moonlighting as a fortune teller, it can still be used in other projects.

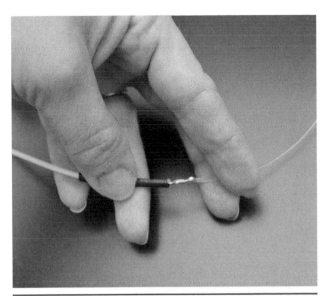

Figure 1-25   Solder the wires, and cover them with heat shrink.

Step 6: Connect Prepared Wiring to Cootie Catcher

Holding the exposed ends of your newly created hookup wires, you will attach each wire to the conductive tape traces from step 2. Use a sewing needle to poke a hole in one end of the arrow on your cootie catcher, as shown in Figure 1-26. Then push the exposed wire from the underside of the cootie catcher up to the conductive tape tracing. Cover the exposed wire with more conductive tape, as in Figure 1-27. This will

ensure connection and hold your wiring in place. Because the wire is so thin and can break easily, make sure that you push some of the plastic-coated wire through to the top of the cootie catcher as well. Repeat this process for each arrow key. Once you have all wires connected to arrows, secure the hookup wires with clear tape or Scotch tape, and bring all the wires to the center of the backside of your cootie catcher (see Figure 1-28).

Figure 1-27   Hold the wires in place with conductive tape.

Figure 1-26   Use a needle for hole placement.

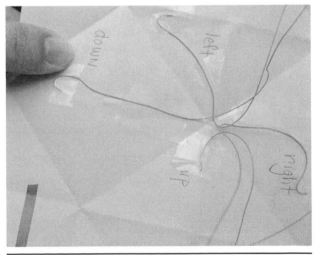

Figure 1-28   Bring all the wires to center, and secure them with clear tape.

Step 7: Prepare "w" and "a" Connections

Connect the wire labeled "w" with conductive tape, as shown in Figure 1-29, and repeat for the "a" wire. Use nonconductive tape to secure the wire to the paper, and bring all the wires to the center of your cootie catcher and refold it so that your project looks like Figure 1-30.

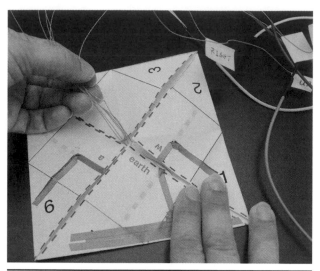

Figure 1-30 Bring all the wires to center, and refold the cootie catcher.

Step 8: Create an EARTH Connection

Grab some tinfoil from your kitchen, and wrap all the wires together as shown in Figure 1-31. Then use conductive fabric tape to wrap around the aluminum foil and up to the tape trace you created as your EARTH input in step 3 (see Figure 1-32). This EARTH connection will complete the circuit for your key presses

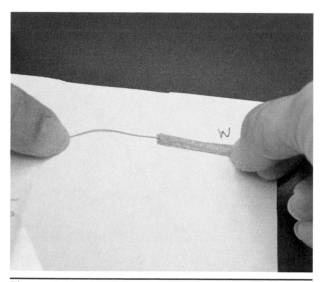

Figure 1-29 Apply hookup wire to the "w" and "a" connections.

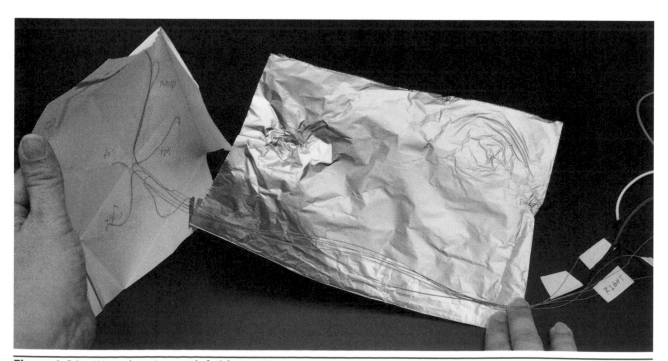

Figure 1-31 Wrap the wires with foil for EARTH connection.

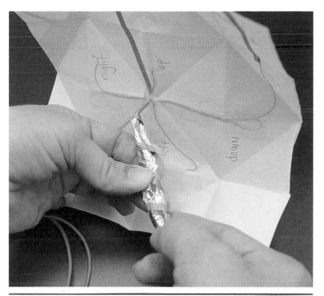

Figure 1-32 Wrap the EARTH connection with conductive tape.

as you open and close the cootie catcher. However, it will also complete the circuit when you tell fortunes, but you will have to be holding EARTH with your hand to make that connection. This is why we are wrapping a long length of these wires with foil—to ensure that we complete that circuit for telling fortunes.

Step 9: Connect to the Makey Makey

Now that your cootie catcher is assembled, clip the alligator clips to the appropriate arrow keys on the Makey Makey, and attach a jumper wire to the "w" pin and another jumper wire to the "a" pin in the header on the back of the Makey Makey. Clip your respective alligator clips to each jumper wire as in Figure 1-33.

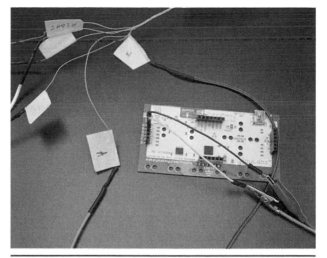

Figure 1- 33 Connect to the Makey Makey "w" and "a" pins.

Connect each arrow wire to the arrow inputs on the Makey Makey, as shown in Figure 1-34, making sure to match your labeled wires with the correct inputs on your controller.

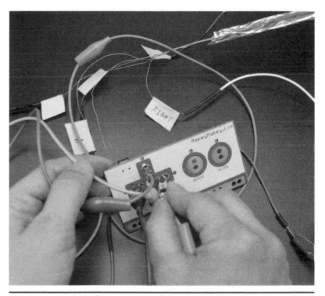

Figure 1- 34 Connect to the arrow inputs.

Step 10: Test Connections

Plug your Makey Makey into the computer, and test all of your connections to make sure that your wires are soldered properly.

Step 11: Tell Fortunes!

Find your favorite (or maybe least favorite) minion, and begin telling fortunes with your cootie catcher! See a video of this project in action at https://colleengraves.org/makey-makey-evil-genius-book/.

Taking It Further

While you are making this project, you should begin to realize how easy it is to create longer and longer conductive connections for Makey Makey. You can extend any connection by connecting conductive items with aluminum foil, wires, or conductive tape. How big could you go with your Makey Makey connections? Could you make a room-sized game of Simon? Could you make a poster-sized cootie catcher? What other materials would be fun to make extralong connections? This project also dabbles in the basic concepts behind AND gate circuits in which both switches must be pressed for the circuit to be closed and therefore "on." What else could you do with combination presses for Makey Makey? How many banana keys could you make for your banana piano by incorporating AND logic? Check out Liam's 88-Key Banana Piano in *Makey Makey Labz*!

Makey Makey Marble Maze

SOMETIMES AS AN EVIL GENIUS you must challenge your old arch nemesis to a timed maze game. Or maybe you want to trap your nemesis in a maze and pretend that escape is impossible? Either way, this is a superquick and fun project build, which integrates scoring and creating a timer with variables in Scratch, that is sure to dazzle and puzzle your greatest enemy (see Figure 1-35).

Cost: $

Make time: 30 minutes

Skill level: 🍌🍌

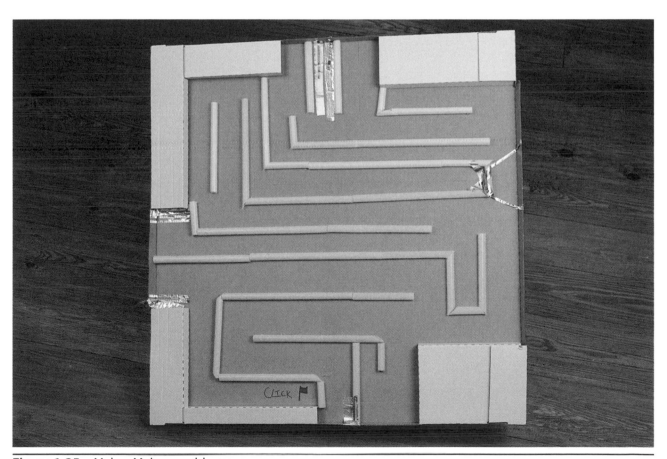

Figure 1-35   Makey Makey marble maze.

Supplies

Materials	Description	Source
Shallow box or box lid	Large, sturdy corrugated cardboard box lid or shallow box with lid removed	Recycling
Steel ball	⅝ inch	Amazon
Grocery supplies	Milkshake straws, aluminum foil	Grocery store
Hot-glue gun and sticks	Low-temperature glue gun	Craft store
Conductive tape	Copper tape with conductive adhesive	Amazon, SparkFun
Chibitronics LED stickers or LEDs	LED circuit stickers or 5- or 10-mm LED	Chibitronics or SparkFun
Computer and access to Scratch	A free visual programming language and online community	scratch.mit.edu

Build a Maze and Switches

Step 1: Find a Big Flat Box and Design a Maze

A pizza box lid or other large flat box will work great for this marble maze. Use a pencil to draw out a complicated maze with many hidden traps and false endings.

Step 2: Build Walls with Straws and Hot Glue

Use hot glue to attach milkshake straws to your box as the walls of your maze. (Milkshake straws are oversized and will help to keep your marble from jumping the walls of your maze.) Cut the ends of straws at an angle to make nice perpendicular corners for your walls.

Step 3: Create Switches at Traps

You can create switches as you create your maze or after all the walls of your maze are built. Any good game should have a hidden point score, a trap that costs you points, and a signal that you've reached the end of the maze. Determine where you want your traps, and let's get started designing switches.

Because the metal marble will complete the circuit, you'll need to place your EARTH and key press very close together. For the "hidden star" switch in Figure 1-36, we actually ran an aluminum foil EARTH and then placed our key press over this EARTH foil. You can do this by creating an insulating layer with regular clear tape to keep the switches from constantly connecting. Make sure to test as you build so that your key press doesn't press over and over. You can also insulate with paper, masking tape, or any nonconductive material you have on hand. Get experimental with this and incorporate your ideas to coordinate your maze design. To achieve the points of the hidden star, the maze runner makes the connection by getting the metal ball to touch both pieces of copper tape at one time, as shown in Figure 1-36. To keep the maze path clear, run the copper tape trace to the outer edge of the box, making sure as you do so that the two copper tape traces are far apart. If your tape traces get too close, it might cause a key press as the ball travels other parts of the maze. In Figure 1-36, the copper tape trace connects to the foil to create more surface contact for the "hidden star" switch.

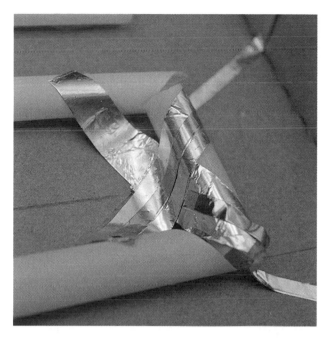

Figure 1-37 "Dead end" switch.

The "finish line" switch will let Scratch know that the racer has reached the end of the maze (see Figure 1-38). It is similar in construction to the "dead end" switch. Make sure that the foil is very close together but not touching. You'll want to make a long runway for your ball to cross. Otherwise, the Makey Makey may not register the ball as a key press.

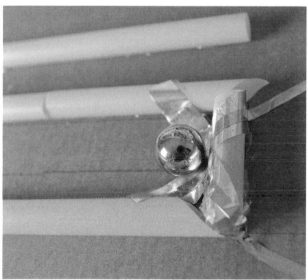

Figure 1-36 "Hidden star" switch.

The "dead end" switch, as shown in Figure 1-37, will let racers know that they've gone the wrong way and deduct points accordingly. This switch is a little more straightforward. The foil on the cardboard will be connected to an EARTH input on the Makey Makey, and the foil on the straw will connect to a key press. Use copper tape to extend the tape traces (the tape is taking the place of the wiring) to the back of your maze. (This step will be explained in step 6.)

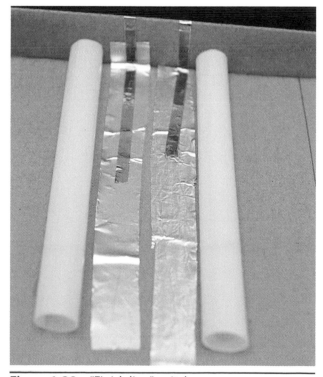

Figure 1-38 "Finish line" switch.

Step 4: Test Switches/Debug

Test these switches by hooking up an alligator clip to one foil lead and the other end to the EARTH input on your Makey Makey. Hook a second alligator clip to the other foil lead on your other trace, and clip it to any key input on your Makey Makey. Roll your metal marble over the switch to test connectivity. If it doesn't make a connection, here are a few debugging tips. First, will it work if the marble is placed across the connections? If your switch is working when the marble is slow but doesn't register a key press when the marble is moving fast, then you need to extend the length of time that the marble will cross the connections and complete the circuit. Because metal and foil are not flexible, the ball may only create the contact during one threshold of the sampling rate on the Makey Makey. If this happens, the controller might think that this interaction is just noise and not an actual key press. Since the metal ball touching the foil traces doesn't cover enough surface area, the signal might glitch, and you need the ball to cross the threshold for more samples on the Makey Makey. If this is so, increasing the length of time the ball crosses the connections will help. You can also try plugging your computer power into the wall to increase the difference for EARTH (not plugged in, EARTH is only the computer battery, but plugged in, the house becomes part of the grounding for EARTH.) Lastly, sometimes it helps to try switching which foil lead is EARTH and which lead is the key press.

Step 5: Power LEDs

If you don't want to power LEDS, you can skip this step, but since our box had two traps where the ball could actually get stuck, we thought it would be quirky and fun to have a LED light up to signal danger, as shown in Figure 1-39. This Chibitronics LED will only light up when the ball makes the connection across the LED

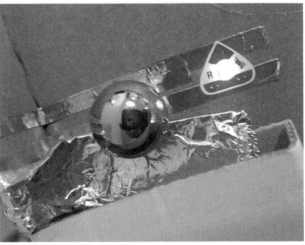

Figure 1-39   LED placement with switch.

traces and the foil switch. To wire this up, you'll need to lay down two copper tape traces for the positive and negative leads on your LED, as in Figure 1-39. For the switch, you'll need a small piece of foil that you'll continue wiring with copper tape traces to the back of your box. Following the template in Figure 1-40, the top tape trace will be attached to your EARTH input, therefore grounding your LED. To power your LED, the second tape trace will lead to the key press input of your choice on your Makey Makey. Lastly, to make the LED light up only during this specific key press, you'll attach a copper tape trace to the aluminum foil switch,

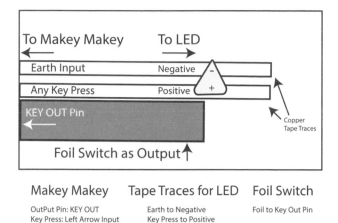

Figure 1-40 Template for wiring output with switch.

and carry the copper tape to the back of the box, where you can attach a jumper wire. Since we have two outputs, we used a jumper wire to connect these outputs on the back of our maze, as shown in Figure 1-41. We insulated our positive and negative copper tape traces with a

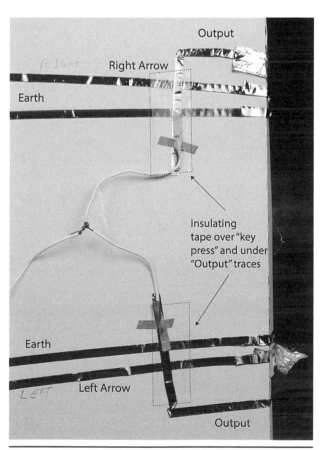

Figure 1-41 Output wiring extended on back of maze.

piece of regular nonconductive clear tape. Then we secured the jumper wires with some fabric tape from the Makey Makey Booster Kit. The jumper wires are twisted together as in Figure 1-41 and then connected to the KEY OUT input pin on the back of the Makey Makey, as in Figure 1-42.

Figure 1-42 Output wiring to Makey Makey.

Optional Tinkering: Chibitronics and LED Powered with Makey Makey

This is actually a pretty strange hack that might seem like it shouldn't really work. The cool thing is that it will only light up with this key press instead of lighting up with any and all key presses (as we wired up our motor in Project 1 and the Morse code machine in Project 8). What is happening here is that the Makey Makey OUTPUT pin is actually always sending a little bit of a signal out, but the LED is only fully powered when the connection between the KEY OUT and OUTPUT pins are bridged. You can continue your conductive tape trace to the other side of the EARTH trace, add another LED if you'd like to tinker with this output trick, and see the difference by placing a metal washer across the different tape traces (see Figures 1-43 and 1-44). You can also try this trick with a regular LED, but you will want to use at least a 100-ohm resistor (also shown in Figure 1-43).

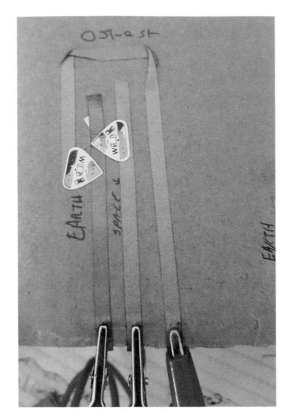

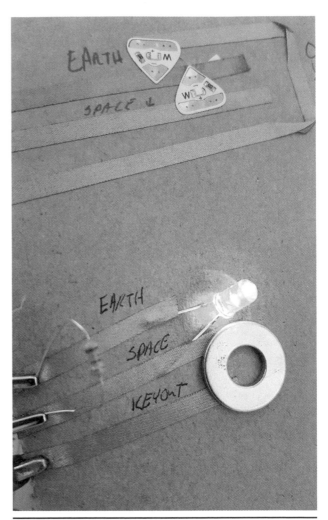

Figure 1-44 LED output tinkering.

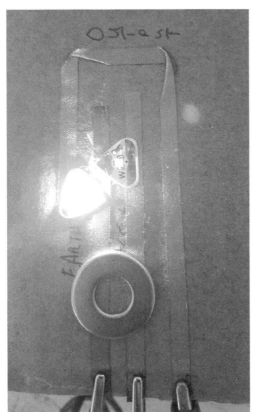

Figure 1-43 Chibitronic output tinkering.

Step 1: Extend Wiring

Use copper tape to extend the conductive traces to the center of the back of your box, as in Figure 1-45. In this way, you'll be able to tuck away your Makey Makey and wiring in the original packaging and keep wires out of the way of the minions' little fingers as they are playing this game. Make sure to write the names of each key press by copper tape traces. Plug alligator clips to each tape trace, as shown in Figure 1-46. Secure the wires of your alligator clips to the maze box with pieces of regular nonconductive clear tape. This tape will insulate your clips and keep them in place. Your maze and switches will be in completely different

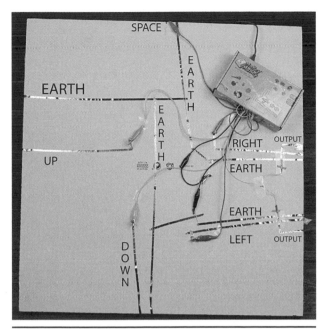

Figure 1-45 Back view of marble maze (labeled).

Figure 1-47 Hiding the Makey Makey in the box.

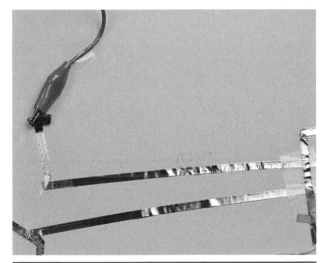

Figure 1-46 Close-up of alligator clip to copper tape trace.

clipped Makey Makey in the box, and hot glue the Makey Makey box to the bottom of your maze box. Now you are ready to program your maze with Scratch.

Program in Scratch

Step 1: Key Press Scripts

Take a picture of your maze to include as the backdrop in Scratch using the "Camera" feature. (Make sure to allow camera and microphone settings for Adobe Flash Player.) Program key presses as you did in Project 2 with the "When key press" and "Play sound" blocks. You can add a "Go to x: y:" script if you want the Scratch interface to show where the ball is in your maze. If you want your Scratch sprite to go to the ball's placement in your game, drag your sprite to a switch pictured in the backdrop. Now use the "go to" block located in the "Motions" palette, and connect it with the key press you want triggered with that switch. Move your sprite for each switch, and notice how the x,y coordinates change as you move your sprite around the stage. You will need a different "When _____ key

places than ours, but see the template in Figure 1-45 for a suggested circuitry layout.

Step 2: Hook Up the Makey Makey

Hook up all the alligator clips to the Makey Makey, and cut a hole for your USB as in Figure 1-47 so that you can easily plug your Makey Makey into the computer. Then hide your

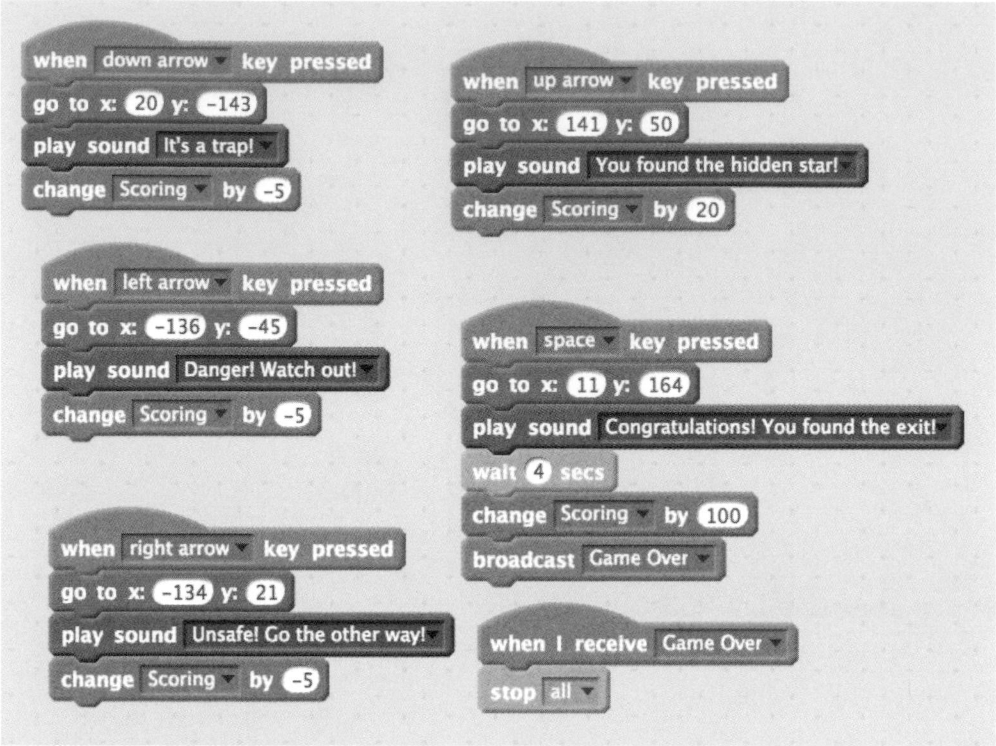

Figure 1-48 Key presses programmed in Scratch.

pressed" block for each switch, along with a "Go to x: y:" block and a "Play sound" block. See the full scripts for key presses in Figure 1-48. You'll see that we've added a scoring variable, but we'll take care of that in the next step.

Step 2: Add Variables

Variables, which you can create in the "Data" palette, will allow switches to track scoring and create a timer for your game. In Scratch, variables allow you to store values and can help to make your games infinitely complex. Open the "Data" palette, and click "Make a variable." We only have one sprite for this game, but notice that you can create variables "For all sprites" or "For this sprite only." Choose "For all sprites," click "Okay," and name your first variable "Scoring." While you are here, go ahead and create "minutes," "ms," and "seconds." Your "Data" menu should now look like Figure 1-49.

Now you can add "Change scoring by ____" to each key press. Remember to deduct points for traps and reward points for finding the "hidden star" switch and, of course, a big point payout for completing the game. (Refer back to Figure 1-48.)

Add a Timer

Step 1: Set the Timer to Start

You need to add your variables to "Show" and set each variable back to zero under the event "When flag clicked." This will reset scoring to zero and can create a timer. First, you'll want to set each variable to "Show" underneath your sprite's initial "Go to x: y:" block on the "When flag clicked" script. Next, drag a "Set ____ to 0" block for each variable to the script, as in Figure 1-50.

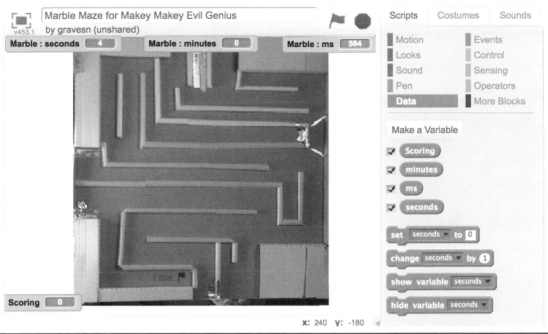

Figure 1-49 Adding variables.

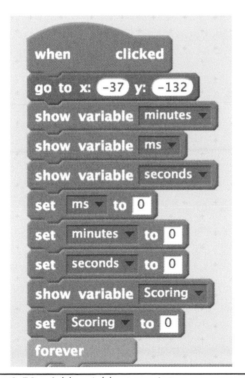

Figure 1-50 Add variables to scripts.

Step 4: Make Seconds

You aren't done yet! To make your timer work, you'll have to add a conditional statement called a "forever" loop to your "When flag clicked" script. This will keep your time going forever until you stop the game by clicking the "Stop" sign. To make the seconds function like a second in Scratch, add a "Change seconds by" block inside the "forever" loop and a "Wait 1 secs" block underneath that as in Figure 1-51. You'll find the addition operation under the "Operators" palette. Drag it into the circle located inside the "Change seconds by" block. When the circle is highlighted, it will click into place. Your timer will now create seconds once you click the green flag. Try it to check it out!

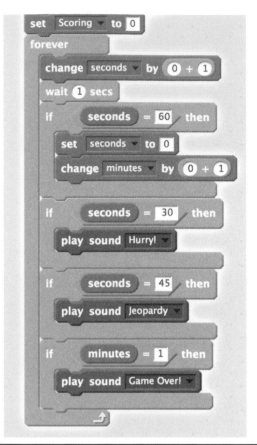

Figure 1-51   Adding timed sounds.

Step 5: Make Minutes

To make seconds turn into minutes, you'll have to add a conditional "if/then" statement inside your "forever" loop. This will tell your game that if the timer hits 60 seconds, it should change the minute variable by one. Drag the operation block "☐ = ☐" into the hexagon shape of your "if/then" statement. Click back to your "Data" menu to add the variable "seconds," and type "60" on the other side of the equals sign. Drag the "Set seconds to 0" and "Change seconds" blocks into the "if/then" statement. Click the dropdown arrow to make your block read "Change minutes by," and go back to your "Operation" menu to add "0 + 1" as the operation, as shown in Figure 1-51.

Step 6: Add Timed Sounds

If you'd like a sound at a certain time in your game to get your maze runner moving more quickly, you just have to add an "if/then" statement and the same hexagon-shaped operation block "☐ = ☐" into the "if/then" statement. Set your operation to read when you want your sound effect to start. In Figure 1-51, we set a sound effect at 30 seconds and 45 seconds and ended our game at 1 minute. Place a "Play sound" block inside each "if/then" statement, and record the sound effect you want to play at each timed instance. We also added a "Broadcast message" block to the "When space key pressed" script that would signal the end of the game if your ball makes it to the end of the maze. Make sure that you add a "When message received" event block to your work area as well and a "Stop all" script to officially end your game (see Figure 1-48).

Debug the Timer

If your timer is acting quirky, you may want to change your variable from the "=" operator to a ">" operator. Do this only if Scratch isn't changing time the way you want it to. Programmers use the ">=" variable in case the program isn't catching the number at the equal sign; it will eventually catch the number that is greater than the variable in the comparison. Since Scratch doesn't have a ">=" operator, you can use the ">" operator if you need to debug your timer (see Figure 1-52). However, you won't want to use this comparison in your timed sounds or they will all sound off at once! In that case, you'll want to start the message between times, for example, if seconds is greater than 30 but less than 34.

Now you are ready to play your maze game!

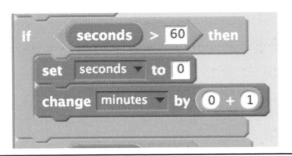

Figure 1-52   Greater-than comparison.

Taking It Further

Creating variables in Scratch and using operations to control your game or animations

will make your creations surprisingly more complex very quickly! What other iterations could you make of this marble maze? Could you create a complex evil plot about your nemesis told throughout Scratch as the ball travels the cardboard pathway? Can you tinker with physical computing and the idea of the ball traveling from the maze seemingly into the computer screen and back out again?

Makey Makey Arcade Coin Slot

EVEN AN EVIL GENIUS NEEDS cash flow, and in this project, you are going to learn how to quickly create a coin slot box that you can add to almost any arcade project (see Figure 1-53). It's a great way to raise funds quickly for that Makey Makey activated laser you so desperately want.

Cost: $

Make time: 30 minutes

Skill level: 🍌🍌

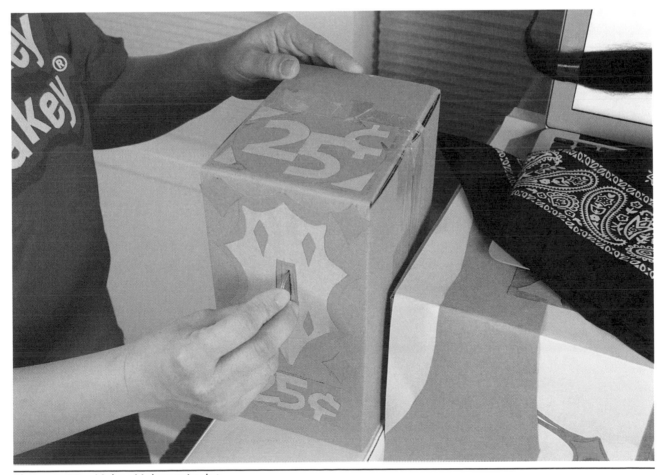

Figure 1-53   Makey Makey coin slot.

Supplies

Materials	Description	Source
Cardboard box	Small box about 6 to 9 inches in length, about 6 inches wide, and 4 inches high	Recycling
Paperclip	Large metal paper clip	Junk drawer or office supply store
Conductive tape	Copper tap with the conductive adhesive or fabric tape from Makey Makey Inventor Kit	Amazon or Joylabz
Hookup wire	Elenco 6 Color Hookup No. 22 AWG Solid Wire Kit or stripped Ethernet cable	Amazon
Tools	Box cutter, craft knife, pliers, duct tape	Toolbox or hardware store
Computer and access to Scratch	A free visual programming language and online community	scratch.mit.edu

Step 1: Create a Slot

Take your small box, leave the top of the box open, and turn it to the side. This will allow you to access the inside of the box later to remove coins and adjust the switch, if needed. Create a coin slot in the center of the side of a cardboard box by cutting two parallel 1-inch lines about $\frac{1}{8}$ inch apart. You can adjust the size of the slot to fit any size coin. This one is designed for a U.S. quarter.

Step 2: Z-Shaped Paperclip Swing Switch

Pry a large paperclip open, separating the small and large sides as shown in Figure 1-54. Unbend the curved ends so that the paperclip forms the letter Z. Bend the paperclip back toward the center to let the top and bottom of the Z rest on your work surface with the center elevated. Cut a $\frac{3}{4}$-inch strip of corrugated cardboard about $1\frac{1}{2}$ inches long. Slide the large end of the paperclip into the corrugated folds of the cardboard about $\frac{3}{4}$ inch from the bottom of the strip, as shown in Figure 1-55. Bend the clip just slightly up as it passes through the cardboard to hold it in place. The clip should pivot freely.

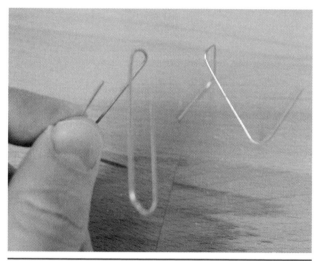

Figure 1-54   Z clip.

Step 3: Placement and Fine Tuning

Place the cardboard and paperclip in the box just above the top of the slot, as shown in Figure 1-56. Align the smaller part of the clip so that it is just above the center of the slot. Hold the strip in place, and test the alignment by pushing a coin into the slot. You will notice that the Z clip pivots. Remember how you bent the large end of the paper clip to hold it in place in the cardboard? You will need to bend that piece of the paper clip so that when a coin pivots the clip, this section will touch the side of the box.

Figure 1-55   Clip in cardboard.

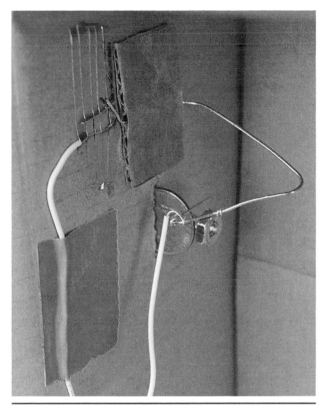

Figure 1-56   Coin slot in action.

Once you are sure of the placement of where the paperclip touches the box, as shown in Figure 1-56, mark the contact area and position of the switch. Remove the switch, and strip about ¾ inches of insulation off of a 3-foot length of hookup wire. Curl the end, and position it underneath or near the contact area. Use several strips of copper tape to make a contact area and to cover the end of the exposed hookup wire. Secure the wire to the side of the box with duct tape.

Before installing the switch, weight must be added to the bottom of the paperclip to ensure that it returns to its resting position. You can slip a washer or small bolt onto the corner of

the bottom of the switch. Strip 1 inch of a 3-foot length of hookup wire. Use the exposed hookup wire to secure the weight onto the bottom of the switch. Make sure that you wrap it tightly so that you have a good connection to the paperclip. Hot glue the cardboard switch in place, and test how much movement the wires will need before securing them to the box. Test the switch several times before you tape the box closed to ensure that your switch is reliable. Remember, you will need to periodically adjust the switch and remove the coins, so be mindful of the need to open the box when you decorate it.

Step 4: Use the Coin Slot with Existing Games

Many online games already require you to press a specific key to start the game. If you are using one of these types of games, route the wire from the coin box to the corresponding key input on the Makey Makey. If that key is not listed, you can remap the keys on your Makey Makey by going to http://makeymakey.com/remap/ (full instructions on remapping the Makey Makey classic are also given in Project 12).

In many Scratch games, the action is started by pressing the green flag. In some instances, you can look at the scripts created for the sprites in the game and switch out all the existing "When green flag clicked" blocks with "When space clicked" or another key press of your choice. After exchanging blocks, create a new sprite with a costume that has the text "Insert coin to start." For this sprite, drag a "When green flag clicked" block to the "Scripts" area, and follow it with a

"Show" block, as shown in Figure 1-57. Add a "When space clicked" block, and follow it with a "Hide" block from the "Looks" menu. Now when the green flag is clicked, the game will start with a message to insert a coin, and all former actions that were triggered by the green flag click will be activated by pressing the space key.

Taking It Further

This is a great way to turn your own games into money makers! How could you change the code to create a 50-cent or 75-cent game? Is there a way to sort out pennies and other coins?

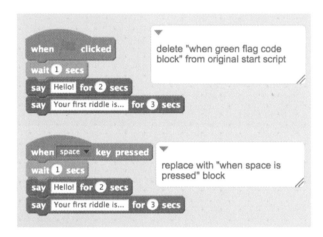

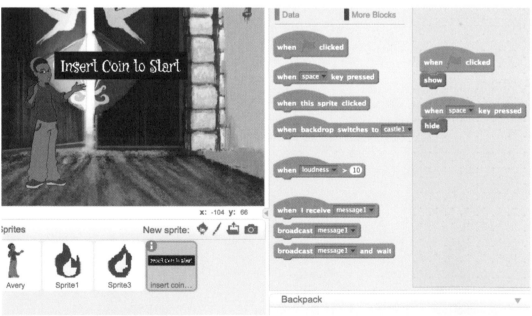

Figure 1-57   Programming for the coin slot.

Arcade-Style Fortune Teller

As AN EVIL GENIUS, controlling the future is always on my list of evil tasks. If controlling the future is out of the question, what if an evil genius could control his or her minions' future. In this project you will make your own fortune teller with Makey Makey by creating a game in Scratch that will randomize fortunes.

Use your fortune teller to serve up backhanded compliments and less than fortunate fortunes (see Figure 1-58).

Cost: $

Make time: 2 hours

Skill level: 🍌🍌🍌

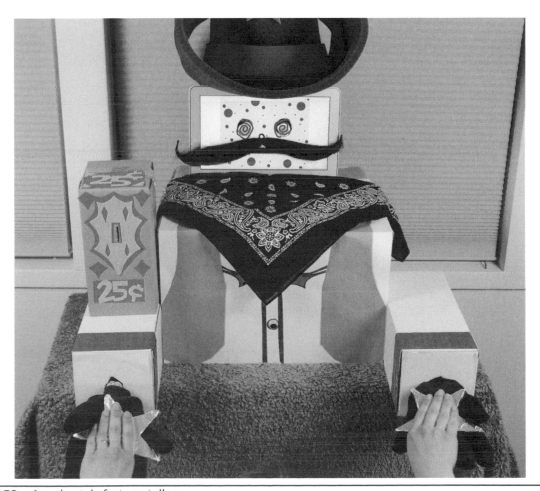

Figure 1-58 Arcade-style fortune teller.

Supplies

Materials	Description	Source
Variety of cardboard boxes	16- by 10- by 10-inch heavy cardboard box, two approximately 5- by 5- by 12-inch boxes	Recycling
Hookup wire or old Ethernet patch cable	Elenco 6 Color Hookup No. 22 AWG Solid Wire Kit or stripped Ethernet cable	Recycling or Amazon
Craft supplies	Construction paper, aluminum foil, scissors, box cutter, clear tape, craft glue, hot-glue gun and sticks, ⅜- or ½-inch wood rod	Craft store
Gloves	Brown or black knit gloves	Dollar store or recycling
Computer and access to Scratch	A free visual programming language and online community	scratch.mit.edu

Build Character and Arcade Box

Step 1: Create a Character

Decide on what type of character you want your fortune teller to be. We chose to make a cowboy fortune teller that would dispense cowboy wisdom and wit. Although this is a temporary cardboard arcade machine, the boxes must be strong enough to hold a laptop or small screen.

For the base you will need a 16- by 10- by 10-inch heavy cardboard box. The fortune teller will activate when a coin is dropped in a slot inside a box alongside of him. For instructions on how to build a coin drop for an arcade game, see Project 4.

Step 2: Make Arms with Conductive Hands

To create arms for your fortune teller, you will need two long, skinny boxes about 5 by 5 by 12 inches. Cut two pieces of ½- or ⅜-inch dowel about 12 inches long. Push the end of the rod into the middle or ring finger of a glove. Use batting, fabric scraps, or tissue to stuff the glove. For your Makey Makey switch, you'll cut a cardboard star to use as a badge and cover it with foil. (Alternatively, you can use a metal sheriff's badge from a costume and pin it to the glove, but test it with Makey Makey to make sure that it isn't coated and therefore nonconductive.) Cut a 3-foot length of hookup wire, and strip the end so that you can attach it to the foil on the star (or wrap it around the pin of the badge). Push the other end of the wire into the palm of the glove, and run it out the opening at the base of the glove. Wrap a rubber band around the base of the glove to secure it to the dowel. On the small side of the box make about a 1-inch crisscross cut with a box cutter, and shove the wire and dowel into the box, hiding the rubber band and cinched glove as shown in Figure 1-59. Repeat the process for the other hand.

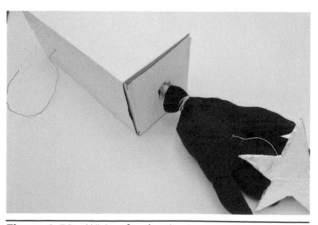

Figure 1-59   Wiring for the tin star.

Step 3: Wire Arms

Turn the base box so that the open end becomes the back. This will allow you easy access to the wires and Makey Makey. On the 10- by 10-inch side place a mark in the middle along the bottom at 5 inches on both sides. Align the arm box next to this mark, and make a hole to route the wire from the arm hole into the main body. After the wire is placed into the main body of the box, hot glue the box in place as shown in Figure 1-58, and repeat the step for the opposite side. Attach an alligator clip to the wire coming from each hand, and route one to an EARTH input and the other to the SPACE input on the Makey Makey (see Figure 1-59).When both stars are touched, the circuit will complete, and "Crazy-Eyed Clyde" will dispense your fortune.

Step 4: Reset Switch and Connecting Coin Slot

Secure your coin slot to Clyde with some hot glue, and route the hookup wires into the base box. Connect alligator clips to the wires, and route one to the SPACE key on the Makey Makey and the other to EARTH. Make a small pressure switch like the one made in step 4 of Project 1. Use two 1-foot lengths of hookup wire for your switch. Place the finished switch on the back of one of Clyde's arms or in an inconspicuous spot, and route the wires into the main box. Attach alligator clips to both wires, and clip the end of one to EARTH and the other to the down arrow input.

Program with Scratch

Step 1: Variables in Scratch

Create a new program in Scratch. In this project, variables are used to trigger and reset the actions of the fortune teller. To set up a variable

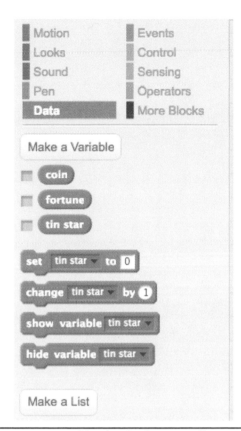

Figure 1-60 Variables for the fortune teller.

in Scratch, click the "Scripts" tab, select the "Data" palette, and click the "Make a variable" button. Type in the variable name, checkmark the "Available for all sprites" button, and click "OK." Create the following list of variables: "coin," "fortune," and "tin star," as shown in Figure 1-60.

Step 2: Animate Backdrops

Crazy-Eyed Clyde will need six backdrops to change his mental state and to control scripts in the program. To create Clyde's face, click on the "Stage" area located below the "Preview" window. Once you do this, the tabs at the top switch to "Scripts, "Backdrops," and "Sounds." Go to the "Backdrops" tab to paint a background. Select the circle shape from the menu, place the mouse near where you want to position the eye, then click and drag until the eye is the size and shape you desire, and then release the mouse button. After the initial release, you can rotate, resize, or position the shape using the white boxes that outline your shape. For more detail, change the color of the circle and add a solid shape to create the iris of the eye, as shown in Figure 1-61. To create the mouth, use a narrow solid black rectangle. Get as detailed as you desire, or keep it simple if you are ready to move on.

Change the name of the backdrop to "Cowboy1" by clicking in the box just above the upper-left corner of the backdrop. To make Clyde's eyes look a little crazed, right-click on the backdrop and choose "Duplicate." Name this backdrop "Cowboy2," and make crazy eyes with the circle shape by adding rings around them and changing the color to red. Right-click the backdrop and duplicate it to create "Cowboy3." For this face, use the paint bucket to fill the background with a light blue. You will use the next few backgrounds to make Clyde look uneasy and create a strobe effect. Select

the paintbrush tool, choose the color orange, and then use the slider to adjust the line width. Click on the backdrop to create random dots around Clyde's face. Duplicate the background twice to create "Cowboy4" and "Cowboy5." Leave Cowboy4 alone, but click on Cowboy5 and use the fill bucket to change the background to orange. Select the paintbrush tool and change the line width to create light blue dots as shown in the examples in Figure 1-61. Right-click on the "Cowboy5" background icon, and choose "Duplicate." Repeat the process for the "Cowboy6" background, but use light green for the background and red for the dots. You can also add more dots and change the size, which will give the background an animated effect. To create the final background, right-click and duplicate to create "Cowboy7." This background will signal the end of the program, so click the "Clear" button to remove all the items from the background. See Figure 1-61 for examples.

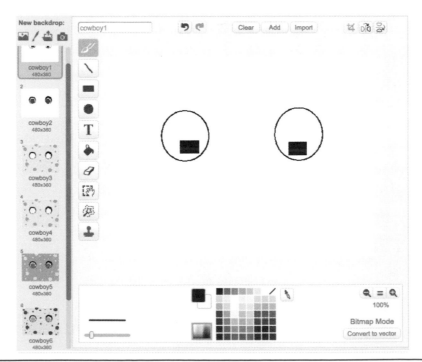

Figure 1-61   Backdrops for Clyde.

Step 3: Animate Eyes and Mouth

It's now time to create the sprites we will use to animate Crazy-Eyed Clyde's eyes and mouth. Click on the "Paint new sprite" icon located in the "Sprites" menu. Use the paintbrush tool to create a red spiral just slightly larger than the size of the eye (see Figure 1-62). If you can't see the sprite in the viewing window, right-click on the sprite and choose "Show." Once you have the right size spiral, right-click and duplicate to create "Sprite2." You can right-click on the sprites when finished and choose "Hide" as needed.

Animating the mouth requires us to create a new sprite with multiple costumes. Make sure that you are on the "Cowboy1" backdrop before clicking the paintbrush in the "New Sprite" menu to paint your own sprite. Notice that when you are painting a new sprite, you are in the "Costumes" tab for this sprite. You can create multiple costumes here that will aide in the animation of your game. First, draw a

rectangular shape to represent a closed mouth, and position it on the backdrop where you want the mouth located. Right-click on the "Costume1" icon, and choose "Duplicate." Select the rectangle tool, and choose the outline option instead of the solid shape. You are going to make the mouth appear open by drawing a rectangular box that overlaps the top of the rectangle a little more than the bottom, as shown in Figure 1-63. Use the "Line" tool to divide the rectangle into teeth. Use the "Fill with color" tool to make the teeth white, and be sure to black out a few. Once you're satisfied, right-click on "Costume2" and duplicate it to create "Costume3." You have a closed mouth and an open mouth, so now you need to create a toothy smile. Use the "Select" tool and place a rectangle around the top row of teeth. Use the down arrow to move the teeth down to cover the black rectangle, leaving "Costume3" with a toothy grin. Examples of each costume are shown in Figure 1-63.

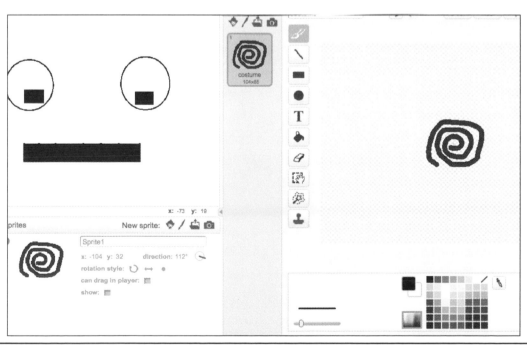

Figure 1-62 Spiral eyes.

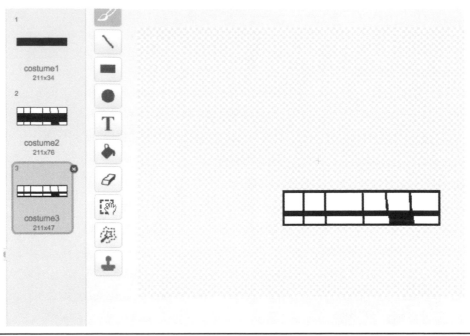

Figure 1-63 Toothy costumes.

Step 4: Record Sounds

To record a sound, click on the "Sounds" tab while you are on the mouth sprite, and press the microphone icon to record a new sound. Press the circular "Record" button, and allow Scratch permission to access your microphone (see Figure 1-64). Once you are finished recording, you can highlight sections of your recording and use the "Edit" or "Effects" menu to alter it. I recommend that you listen to a lot of Pat Buttram, John Wayne, and Slim Pickens to perfect your cowboy accent before recording. You will need to record the following instructions and name them accordingly (or create your own crazy instructions, but make sure that you keep the variables named accordingly: "coin," "crazy," "stars," "hand"):

- **Coin:** "Step right up. Drop a quarter in the slot."

- **Crazy:** "Hear your future. Ol' Crazy-Eyed Clyde will tell it like it is!"

- **Stars:** "Put your hands on those tin stars, and prepare to meet your future."

- **Hand:** "You need somebody to hold your hand or what?"

Make a list of at least five fortunes and nuggets of wisdom before recording. A quick Google search for "cowboy wisdom" and "cowboy sayings" will reveal a plethora of results. Here are a few to get you started (be sure to name your fortunes Cow1, Cow2, and so on):

- **cow1:** "The only thing I can foresee in your future is a bath, maybe two. One for the dirt and another for the grime!"

- **cow2:** "Don't worry about biting off more than you can chew. Your mouth is whole lot bigger than you think!"

- **cow3:** "Never ask anyone how stupid they are or they will turn around and show you."

See Figure 1-64.

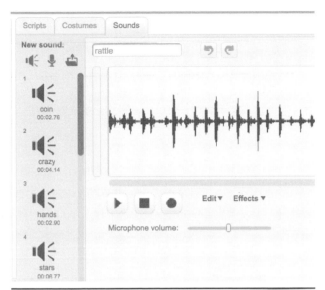

Figure 1-64 Sounds tab in Scratch.

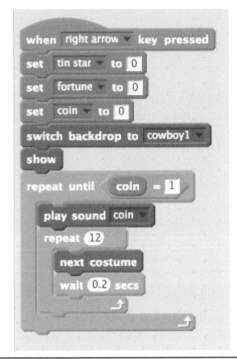

Figure 1-65 "Repeat" block.

Step 5: Reset/Start Button and Resting State

It's time to give all those sprites, backgrounds, and costumes some instructions. Select the mouth sprite and click on the "Scripts" tab. Go to the "Events" palette, and drag a "When space pressed" block to the scripts area. Press the dropdown arrow to change the key press to "right arrow." To set the value of some of the variables to 0, click the "Data" palette, and drag over three "Set variable to 0" blocks and add them just below the "When right arrow is pressed" block. Use the dropdown arrow on each block to assign each of the blocks to the three variables we created earlier: "tin star," "fortune," and "coin" (shown in Figure 1-65).

We will use backdrops to create our fortune teller's face and mood. Select the "Looks" palette, and drag the "Switch backdrop to cowboy1" block to the scripts area. Because the mouth of the face is animated and doesn't appear in the backdrop, add a "Show" block from the "Looks" palette so that the fortune teller's smile magically appears. The fortune teller needs to engage potential customers and give them some instructions until someone drops a coin in the slot to put it in "fortune

telling" mode. This can be accomplished by clicking on the "Control" palette, and dragging over a "Loop until" block to the end of the script. The "Loop until" block has an empty space where a condition can be added. To add a Boolean operator in this spot, navigate to the "Operators" palette, and then drag in the "□ = □" block. Fill in the first value slot with the variable "coin" located in your "Data" palette. In the second value, enter 1, as shown in Figure 1-65. This will allow the script in the "Repeat until" to repeat instructions until the coin value is changed to 1 by a coin being dropped in the coin slot. To program your reset button, you'll need to add a "stop all" block to a "when down arrow pressed" block.

Step 6: "Repeat" Block

Now that we have a loop created, it is time to add a sound so that Crazy-Eyed Clyde can engage the audience. From the "Scripts" tab, select the "Sound" palette, and drag the "Play sound" block into the "Repeat until coin = 1" block.

Use the arrow to select the "Coin" recording. To animate Clyde's mouth, we will create a loop that moves through the different costumes. Click the "Control" palette, and drag the "Repeat 10 times" block under the "Play sound coin" block. Select the "Looks" palette, and drag the "Next costume" block into the "Repeat 10 times" block. To keep the costume changes from happening too fast, go to the "Control" palette, and drag over the "Wait 1 secs" block and place it below the "Next costume" block. Change the amount of time to 0.2 second. Drag another "Wait 1 secs" block to the scripts area, but place this one after the "Repeat 10 times" block. This will create some silence between when Clyde stops speaking and then starts up again. Use the right arrow key to test to see if Clyde's mouth moves for the same amount of time as he speaks. Increase the amount to "Repeat" if the mouth stops before he finishes speaking. Reduce the amount in the "Repeat" block if his mouth keeps moving after the sound is finished playing (see Figure 1-65).

To provide the audience with some information about the service Clyde provides, you'll want to add another recording. From the "Sound" palette, drag the "Play sound" block into the "Loop until coin = 1" block, placing it after "Wait 1 secs" block. Change the sound to the recording "Crazy," and add a "Loop 10 times" block after it. Place a "Next costume block" from the "Looks" palette into the loop followed by a "Wait 1 secs" block. Adjust the wait time to 0.2 second. Place a "Wait 1 secs" block just below the "Repeat" block, and change the value to 2 seconds. Press the right arrow to test and adjust the amount of time Clyde's mouth needs to move.

Step 7: Debug with "if/then"

You added another recording to the loop, and now the repeat time has increased and made the time to complete the script very long. What if the customer puts a coin in during the beginning or near the middle of the loop? To stop the script in the middle, we can add an "if then" condition block. Drag an "if/then" condition block from the "Control" palette. Use the "□ = □" operator in the condition statement and input the variable "tin star" for the first value and "2" for the second. Drag the "Stop all script" block from the "Control" palette into the "then" location. Use the dropdown arrow and change "All" to "This script" (see Figure 1-66).

Figure 1-66 "If/then."

Step 8: Coin Drop

After the coin switch is activated, the coin variable changes to 1. This event cancels the "Repeat" loop in the preceding step and starts a new script urging the patron to touch the tin stars in Clyde's hands. Drag a "When space is pressed" block from the "Events" palette, changing "space" to "left arrow." To make Clyde appear just a little crazier, drag a "Switch backdrop" block and add it to the script. Use the arrow to select the "Cowboy2" backdrop.

Sounds are another good way to signal a change in Clyde's fortune telling state. Select the "Sounds" palette, and add a "Play sound" block to the script. Click the down arrow, and select "Record." Instead of recording, click the speaker icon, and add the "Rattle" sound from the library.

Step 9: Set Variables

From the "Data" palette, drag and attach two "Set □ to 0" blocks to the end of your script. Change the first block to "Set coin to 1" and the second to "Set tin star to 1." To give the preceding script time to complete playing the sounds, drag a "Wait 1 secs" block to the end of the script and change the value to 5 seconds. Depending on how quickly or slowly you speak, you may need to adjust this value.

Both the "tin star" and "coin" variables are changed by the left arrow press. While the coin variable value increase turns the first script off, the "tin star" variable starts the next script and prevents anyone from skipping to their fortune without paying. Add an "If condition is true then" block from the "Control" palette to this script. From the "Operators" palette, you will need to drag a "□ = □" block into the condition location. From the "Data" palette, drag the variable "tin star" into the first value spot. In the second value spot, input 1. To program

Clyde to tell patrons what to do after they have given him a coin, we need to add a "Repeat until" block from the "Control" palette inside this "If condition is true" block. To control the time Clyde repeats these instructions, you will use an operator. You only want to repeat these sounds until the value of the variable "tin star = 2" and add the "□ = □" operator block. Fill in the first location with the variable "tin star" and the second value with 2. Select the "Sound" palette, and drag the "Play sound" block into the repeat loop. Use the down arrow to select the sound "Tin star." Animate Clyde's mouth by dragging a "Repeat 10 times" block below the "Play sound" block. Insert a "Next costume" block from the "Looks" menu, followed by a "Wait 1 secs" block from the "Control" menu. Adjust the time to 0.2 second, and increase the number of loops to about 30. Test and adjust the animation to match the length of sound play.

Step 10: More Debugging with "if/then"

To cut the script short if a patron touches the stars quickly, add an "if/then condition" block from the "Control" palette. Use the "□ = □" operator block in the condition statement and input the value "tin star = 2." Drag the "Stop all script" block from the "Control" palette into the "then" location. Use the dropdown arrow to change the "stop" block from "All" to "This script." Add a "Wait 1 secs" block, and change the value to 2. Start the second recording by adding a "Play sound" block, and select the "Hands" recording. Follow up with a "Repeat 10 times" loop to animate the mouth. Place a "Next costume" block and "Wait 2 secs" block into the loop. To space the recordings, add a "Wait 2 secs" block outside the "Repeat 10 times" loop. Check the order and placement of your script with the example in Figure 1-67.

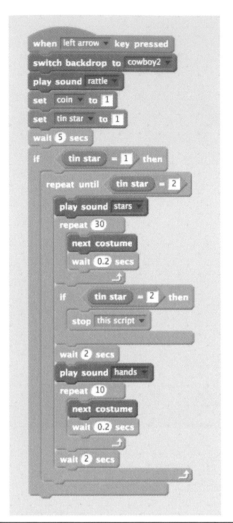

Figure 1-67 Adding an "if/then" conditional statement.

Step 11: Crazy Eyes

Click on the spiral sprite you made for the left eye. Drag the "When space key is pressed" block, and use the arrow key to change it to read "left arrow key." Add a "Show" block from the "Looks" script. We want the crazy eyes to twirl and show until the end of the fortune telling. Add the "Repeat until" script from the "Control" menu to the script. In the spot for an operator, drag the "□ = □" block. Input the variable "fortune" from the "Data" menu, and enter 2 as the second value. Click on the "Motion" palette, and then drag the "Turn 15 degrees" block inside the "Repeat" block. To slow the rotation of the spiral down, add a "Wait 1 secs" block beneath the motion, making sure to change the time to 0.1 second. If the rotation of the spiral is way off center with the eye, click on the "Costumes" menu and click the "Set costume center icon" located in the upper-right corner of the drawing screen. Place the crosshair in the center of the spiral. You may need to move the position of the spiral sprite in the preview window.

Clyde's crazy eyes need to disappear once the game is reset or when the loop starts over. Drag a "When space key is pressed" block to the scripts area, and select the "right arrow" value. To make the crazy eyes disappear on a specific background, drag a "When background switches to" block, and select "Cowboy7." Add the "Hide" block to both of these scripts. To copy the scripts over to the right eye, open the backpack and drag the scripts for the spiral on the left eye into it. Then click the right eye spiral and drag the scripts from the backpack into the scripts area (see Figure 1-68). Be sure to align the center of the spiral. (You can also drag the scripts from the left eye sprite to the right eye sprite to duplicate them into this sprite.)

Step 12: Play Fortunes

To make Clyde start telling random fortunes, click on the mouth sprite, and drag the "When a space key is pressed" block from the "Events" palette to the work area. From the "Data" palette, add a "Change tin star by 1" block followed by a "Change fortune by 1-inch" block. Next, add an "if/then" block from the "Control" palette. Place the operator block "□ = □" in between the "if" and "then" statements. Add the variable "coin" to the first value and the number 1 to the second value. From the "Looks" palette, add a "Switch backdrop" block inside the "if /then" block brackets, and set the value to "Cowboy3." Next, drag two "Play sound until over" blocks from the "Sounds" palette

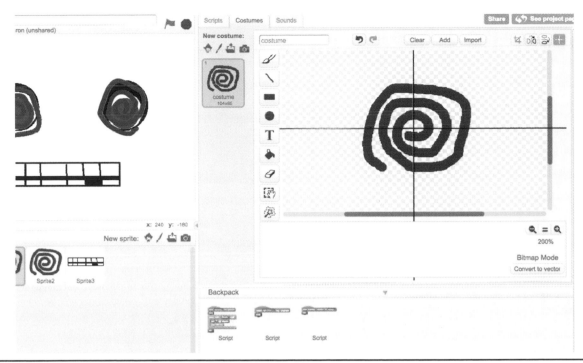

Figure 1-68 Caught in the crosshairs.

to the script area. Press the arrow, and choose "Record," click the speaker icon, and then select the "Horse" sound from the sound library for the first block and the "Horse gallop" sound for the second block. This allows the previous scripts time to finish and alerts the minions that something is about to happen. Add a "Wait for 1 secs" block to the script, and adjust the time if you find the fortune scripts start too soon. From the "Control" palette, add a "Broadcast message1" block. We will use this block to signal the animation script for Clyde's mouth to start.

Add a "Play sound until done" block next, and instead of selecting a sound, drag the "Join hello world" block from the "Operators" palette into the sound block. For the first value, input "Cow," and for the second value, drag the operator "Pick random 1 to 10." Adjust the second number to the number of fortunes you have recorded. This block will now randomly pick one of your recordings to play. Add a "Wait 1 secs" block next and then the "Data" block

"Set fortune to 2." Complete the script with a "Change backdrop to cowboy7" block. This final action will help to reset the program to the beginning (see Figure 1-69).

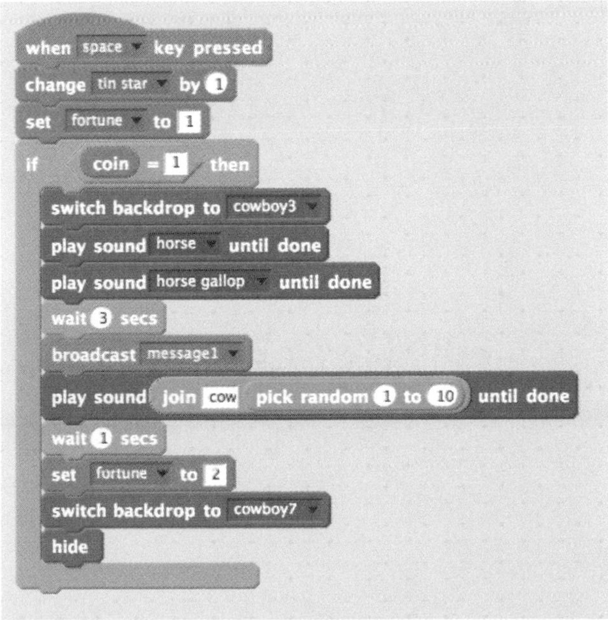

Figure 1-69 Full scripts for space bar.

Step 13: Loose Ends, Variables, and Broadcasts

There are a few things you started in the last step that need finishing touches. When the script from step 12 reaches the block "Switch to backdrop3," you are going to create some code to vary the backdrop images and create a strobing effect to add theatrics. Drag a "When backdrop switches" block from the "Events" palette to the script area to start a new script. Switch the value to "Backdrop3." Attach an "If condition then" block from the "Control" palette, and use the operator block "□ = □". Set the values to "Fortune = 1." Add a "Repeat until" block inside the "if/then" statement. Choose the operator block "□ = □", and set the values to "Fortune = 2." Inside the "Repeat until" block, add the "Wait 1 secs" block and change the value to 0.1 second followed by a "Switch backdrop to Cowboy4." Add a "Wait 1 secs" block, and repeat this step for the backgrounds "Cowboy5" and "Cowboy6," as shown in Figure 1-70.

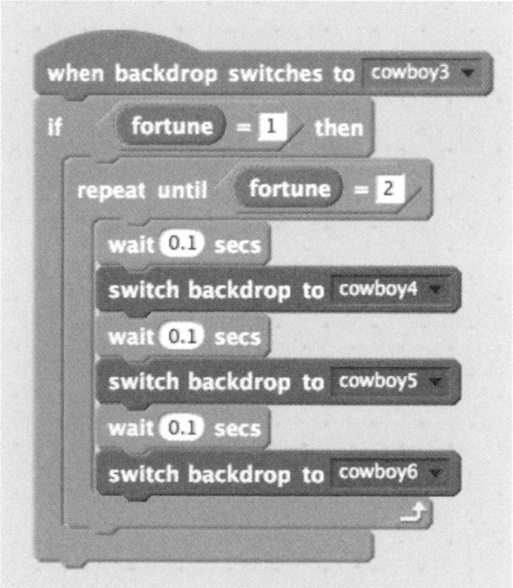

Figure 1-70   Strobing background effect.

In the preceding step we created a broadcast message to start the animation that animates Clyde's mouth, and now it is time to create a script that will receive that message. From the "Events" palette, drag a "When I receive message1" block to the script area. Follow it with an "if/then" block from the "Control" palette. Choose the operator block "□ = □", and set the value for as "Fortune = 1." Add a "Repeat until" block inside the "then" part of your "if" statement. Choose the operator block "□ = □", and set the value to "Fortune = 2." Inside the "Repeat until" block, add a "Next costume" block from the "Looks" palette, and finally add a "Wait 1 secs" block and change the value to 0.1 second. Run your code, and check it with Figure 1-71.

The final key to creating a looping script involves duplicating the first script that is activated by the right arrow key press and triggering it when the background is switched to "Cowboy7." Locate the "Script" group, right-click on it, and click "Duplicate." Replace the "When right arrow key pressed" block with the "When backdrop is switched to" block, and change the value to "Cowboy7." To allow Scratch some time to process all this code, add a

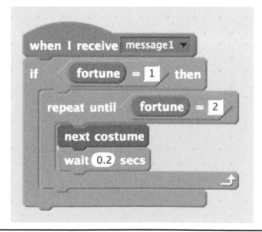

Figure 1-71   "If/then" repeat until.

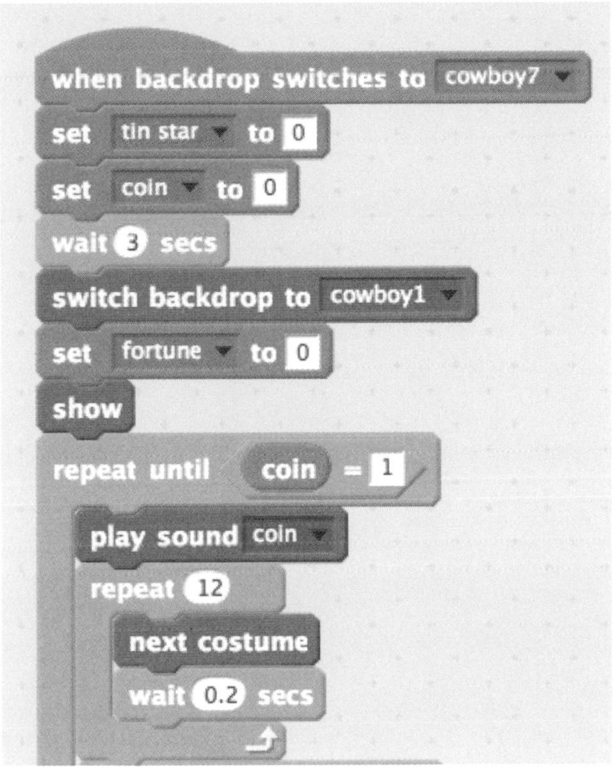

Figure 1-72 "Repeat" looping script.

"Wait 3 secs" block from the "Control" palette just below the "Set coin to 0" block. You will also need to move the "Set fortune to 0" block below the "Switch background to cowboy1" block (see Figure 1-72). Now it's time to tell fortunes!

Step 14: Final Setup

With your game finished, set your laptop on top of the cowboy box, cover your keyboard with a handkerchief, put an oversized cowboy hat on your monitor, and hook up and hide your Makey Makey. Then make sure that you have a bag of quarters ready so that your minions can spend all their money trying to get the finest fortune. See a video of this project in action at https://colleengraves.org/makey-makey-evil -genius-book/.

Taking It Further

What kind of crazy fortune teller can you dream up? What other ways could you get minions to interact with your fortune telling machine? How can you use backdrops and costumes to enhance the animation of your character and the mood of your fortune telling game?

Build Your Own Pinball Machine

THERE IS NOTHING MORE EVIL than a time burglar! A pinball machine is addicting to play and super fun to make. Evil geniuses don't buy pinball kits with tons of fancy flippers, switches, and dials— so you are going to learn to make your own (see Figure 1-73). We will teach you the basics so that you can get started creating your own perfect pinball machine. In this ultimate finale to the "Gaming" section, you'll hone your switch fabrication and woodworking skills and even dabble in 3D design.

Cost: $$$–$$$$

Make time: 30 minutes

Skill level: 🔧🔧🔧🔧

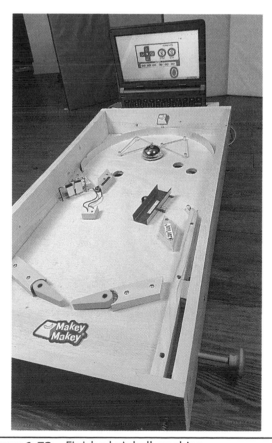

Figure 1-73   Finished pinball machine.

Supplies

Materials	Description	Source
Pinball with Ball Return		
Lumber	8-foot 1- by 2-inch board, 8-foot 1- by 4-inch board, 8-foot 1- by 6-inch board, 24	Hardware store
Handles	Three large wooden knobs	Hardware store
Simple pinball		
Lumber	8-foot 1- by 2-inch board, 8-foot 1- by 4-inch board, 24- by 48- by ½-inch piece of plywood	Hardware store
Nails	1-inch brad nails	Hardware store
Supply List for Both Machines		
Tools	Drill, drill bits, hand or power saw, miter box, scroll saw or coping saw, construction square, rasp, sandpaper, pliers, box cutter, wire stripper, hex key set, hammer	Hardware store or toolbox
Balls	⅝-inch metal ball or ⅝-inch marbles	Amazon or hardware store
Wooden dowel	⅜-inch-diameter wooden dowel or pants rail from wooden hanger	Hardware store or closet
Aluminum strip	36 inches long by ⅛ inch thick by 1 inch wide	Hardware store
Adhesive	Wood glue, general-purpose epoxy	Hardware store
Screws	1-inch No.10 pan head wood screws 1¼-inch No. 10 pan head screws 1¼-inch No. 8 pan head screws ⅝-inch No. 6 screws ½-inch No. 216 screw eyes	Hardware store
Conductive tape and foil	Copper tape, conductive fabric tape, and heavy-duty aluminum foil	Sparkfun or Joylabz
Rod and collars for ball launcher	12 inches of ¼-inch-diameter metal rod, two shaft collars with ¼-inch inside diameter	Hardware store
Washers	⅜-inch ID brass or zinc washers	Hardware store
Optional Supplies		
Bell materials	Metal bell with push button, 2-inch screw or bolt that will fit hole in bell, four nuts and small washers	Hardware and office supply stores
Roller switch	SPDT changeover switch	Radio Shack
Conductive tape	Makey Makey Inventor Kit	Joylabz
Bumper supplies	⅝-inch-height nylon spacers, wide rubber bands, 1-inch No. 10 screws	Hardware and office supply stores
Plexiglas laptop guard	Plexiglas or Lexan sheet	Hardware store
Laptop and access to Scratch	A free visual programming language and online community	scratch.mit.edu
Wire for rails	No. 12 AWG copper wire, clothes hanger wire	Hardware store or closet

In this project, you can choose to construct a basic or more complex pinball machine. Both designs require a handsaw and miter box, coping saw, and drill. Almost all the materials are available at your local hardware store. There are even two different styles of pinball machines to choose from, one that allows the balls to drop through holes and roll to the front and another that holds them in place for scoring and is a simpler build. The accompanying Scratch game will be activated by a pull switch, and a laptop screen will function as the display board. As the ball triggers switches on the machine, sounds will play, and the player's score will be calculated.

Step 1: Cut It Out

Many big-box hardware stores offer 24- by 48-inch pieces of plywood, and if you do not have a table saw, they will make a few cuts for you for a small charge. You will need to get someone to do your bidding and cut a piece of ½-inch plywood to the dimensions in the supplies list for the pinball machine you want to build. Complete the cuts listed on the list in Figure 1-74 for the simple pinball machine and the list in Figure 1-75 if you want to build a machine with a ball return. Many of the obstacles can be created with leftover scraps, so put those aside for now. Continue to step 2 to

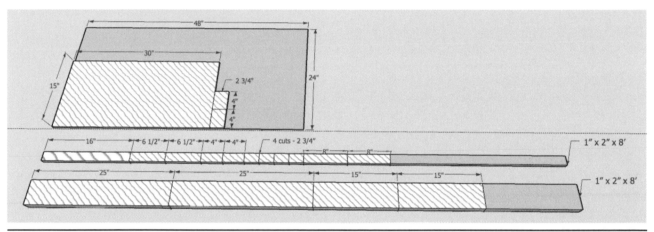

Figure 1-74 Cutlist for simple pinball machine.

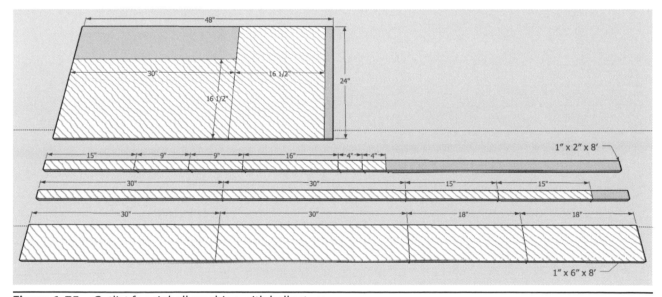

Figure 1-75 Cutlist for pinball machine with ball return.

create the machine with a ball return, or skip to step 3 for the simple pinball machine. Note that it is also possible to use an old wooden drawer with a width of about 15 to 18 inches and adapt the instructions and measurements for the simple pinball machine model. This will require modifying some of the instructions and measurements, but if you want to do this, skip to step 4.

Step 2: Frame and Base Assembly for Pinball Machine with Ball Return

The 1- by 6-inch boards will create the sides, and the 1- by 2-inch board will create a base for the playing surface. For this assembly, it is important to keep in mind that a 1-inch-thick board is actually only ¾ inch thick.

On the 15-inch cuts of the 1- by 2-inch board, use a pencil and square to mark the holes in Figure 1-76. Place a cross in the center of the board at ¾ inch on each of these lines. Use a drill with a ⅛-inch bit to drill a hole through the board. On both of the 30-inch sides, mark at 2, 6½, 15, 21½, and 28 inches, and center the marks at ¾ inch. Continue using a ⅛-inch bit, drill the holes.

On the 8-foot 1- by 6-inch boards, draw a line 1½ inches away from each end. Use a square to draw the mark completely across the 6-inch side of the board. Align the 1- by 2-inch board on the edge of the 1- by 6-inch board between the marks you just made. Fasten them together with a clamp, or have a helper hold them in place. Use a small piece of tape on the drill bit to flag the 1-inch mark, as shown in Figure 1-77. Use the same holes you created earlier, and drill to a depth of 1 inch to create pilot holes in the 1- by 6-inch board. Drilling pilot holes will reduce the risk of the wood splitting and make assembly much faster. Unclamp the boards, and place a small bead of glue on the back of the 1- by 2-inch board. Then reposition and clamp it in place while inserting 1-inch No. 8 screws.

Figure 1-77   Drilling with flag.

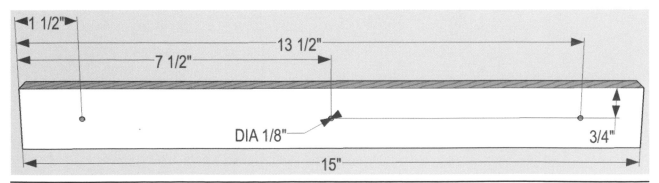

Figure 1-76   Marking holes for the base.

To create the longer sides, align the 1- by 2-inch board with the side of the 1- by 6-inch board, and repeat the previous steps of drilling pilot holes, adding glue, and securing the 1- by 2-inch board in place with screws. Mark one of the long sides with an "L" for left side, and label the other side with an "R" for right side. Lay the left and right sides out with the 1- by 2-inch board facing the inside. Label the ends front and back. Next, we will create holes for the push rods that control the flippers to travel through. On the outer side of the 2- by 6-inch sides, measure and mark from the front $3\frac{7}{8}$ inches and $1\frac{1}{8}$ inches from the bottom, as shown in Figure 1-78. Use your drill to make a $\frac{17}{64}$-inch hole at this location. It may be helpful to start with a $\frac{1}{8}$-inch hole and work your way up to the correct size to avoid splintering.

It is time to drill some pilot holes so that we can place the pieces together to create a frame. Turn the 18-inch sides over so that the side with the 1- by 2-inch board is on the bottom. On the right side, draw a mark $\frac{3}{8}$ inch away from the

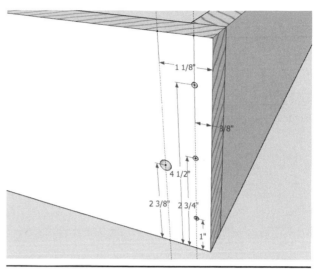

Figure 1-79 Assembly holes in 1- by 6-inch board.

side at the top and bottom. Connect the line, and place crosses at 1, $2\frac{3}{4}$, and $4\frac{1}{2}$ inches, as shown in Figure 1-79. Repeat the process on the left side. After all the marks are made, drill all the way through the wood with a $\frac{1}{8}$-inch bit.

On the front side, we need to create a hole with a $\frac{9}{32}$-inch drill bit for the ball launcher in

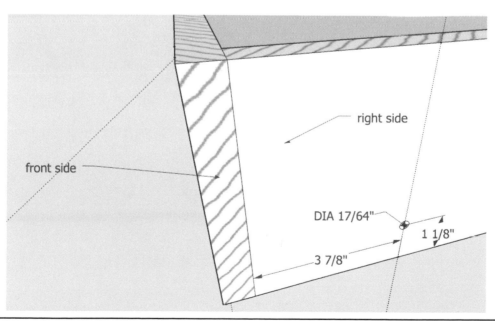

Figure 1-78 Holes for the pushrods.

the lower-right corner. Center the hole 1⅛ inches from the right side and 2⅜ inches from the bottom, as shown in Figure 1-78. After all the pilot holes are drilled, use 1¼-inch No. 10 screws to assemble the base. Use a large carpenter's square to make sure that the base box is square. Drop the plywood into the frame so that it rests on the 1- by 2-inch board. Use a ⅛-inch drill bit to create some pilot holes about every 6 inches around the perimeter of the base. Use some 1-inch No. 10 wood screws with tapered heads to secure the plywood to the base.

Step 3: Box Assembly for Simple Pinball Machine

Cut two 1- by 4-inch boards to 15 inches in length and two more to 25 inches in length. To create some pilot holes for assembly, mark the top of the 15-inch board on the right side ¾ inch from the top and bottom. Place a cross ¾ inch away from the right side. Repeat this step on the left side. You will need to create a hole with a 9/32-inch drill bit for the ball launcher in the lower-right corner. Measure in 1⅛ inches from the right side and ⅜ inch high from the bottom of the 1- by 4-inch board, as shown in Figure 1-80.

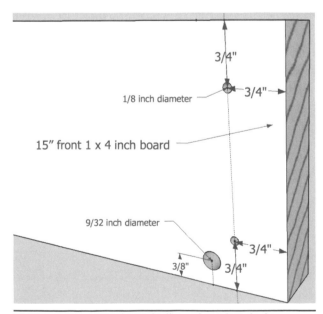

Figure 1-80   Simple pinball machine box holes.

Stand the boards on the ¾-inch side, and use a corner clamp or have a minion hold the 25-inch 1- by 4-inch board at a right angle to the top. With a ⅛-inch drill bit, drill through the wood at a 90-degree angle on these marks. Repeat these steps for the bottom side. Once the two boards are aligned, use the existing pilot holes to drill 1 inch into the longer piece. Secure the two pieces together by using four 1¼-inch No. 10 wood screws. Repeat this step to create the bottom of the box, excluding drilling the hole for the ball launcher.

Use a square to make sure that your box is aligned properly. If it isn't perfect, that's okay; just loosen the screws and adjust it by pushing on the sides. Leave the screws a little loose, and use the plywood base to help you square your box. Lay the plywood down on the box with the sanded side facing down. Create pilot holes around the edges about every 6 inches, and use 1-inch No. 8 wood screws to secure the plywood to the base (see Figure 1-81).

Figure 1-81   Simple pinball machine box assembly.

Step 4: Ball Launcher

You will need the 16 inch long piece of 1- by 2-inch board. Turn the 16-inch piece on its side, and secure it to a table using a clamp. Center marks at 2, 8, and 14 inches. Drill ⅛-inch pilot holes all the way though these pieces. Because we need the screw to sink into the wood further, use a piece of tape to make a flag on a ⅜-inch drill bit to mark ¾ inch, and drill to that depth. Position the 16-inch piece ¾ inch away from the wall, and drill ⅛-inch pilot holes into the plywood. Secure the piece in position with three 1¼-inch No. 10 screws.

Use a clamp to secure the 1½-inch piece to the table so that the cut side is facing up. Mark and drill a ¹⁷⁄₆₄-inch hole ⅜ inch from the bottom, and centered at ⅜ inch from the sides. This piece will serve as a guide for the rod. We were able to purchase a 12-inch piece of ¼-inch steel rod from our local store, but if you have to buy a larger length, use a hacksaw to cut it to a 12-inch length.

To create your pull, you can use large wooden beads with a ¼-inch hole or drawer handles. Mark the center, and drill a ¼-inch hole at least ½ inch into the drawer handle. You may need to use a pair of pliers, a clamp, or a vise to hold the handle upright. Once the hole is drilled, use an epoxy glue to adhere the wooden handle to the metal rod. Once the glue is dry, slide a washer with a ¼-inch hole onto the rod and then a spring followed by another washer. Insert this into the hole, and slide another washer on, followed by a spring and then a set collar (see Figure 1-82). Push on the rod slightly so that the components are under slight tension, and then tighten the setscrew on the collar. Slide the 1½-inch wood guide onto the rod, and position and tighten a set collar on the end of the rod. Pull the rod back all the way, and mark the spot where the inner edge of the collar ends. This will be the spot where the collar will rest against the wooden guide. Secure the guide by placing a 1-inch No. 8 screw through the side of the machine. Once all the parts are tightened, place a ball in the chute and pull back or strike the rod with a quick blow. If your ball does not shoot out onto the playing surface, adjust and tighten any loose parts.

Figure 1-82 Ball launcher.

Step 5: Curve

To get the ball to curve around the playing field, you can use a thin piece of aluminum or cut circular shapes out of wood to direct the ball's path. We used a 1⅛-inch strip of aluminum that was about ⅛ inch thick. Slide the aluminum piece into the ball shooter chute, and gently bend it around the back end of the playing field. Slowly bend the strip around into the desired shape, as shown in the completed pinball machine in Figure 1-73. To hold the strip in place, either drill pilot holes and secure it with small tapered-head screws or with a pan-head screw placed just above the aluminum strip. By tightening the screw, the pan head should overlap the strip and hold it in place. If you drill holes to hold the strip in place, position them at least ¾ inch high to avoid knocking the ball off the path.

Step 6: Flippers

The flippers for this project can be 3D printed or made with some scrap 1- by 2-inch board. To create wooden flippers, mark a 1- by 2-inch board at 3 inches long. Place a washer or something with a 1-inch diameter so that it touches the bottom-right edge of the rectangle, and trace it. On the opposite end, place a washer with a ¾- to ½-inch diameter, and trace it. Connect the top line with a ruler. Mark the center of the larger 1-inch washer, and drill a ⅜-inch hole. Be sure to drill a pilot hole and work up to the correct size, keeping the drill at a 90-degree angle. Repeat the process, only this time place the larger washer on the left side. This will give you two very straight playing surfaces in case your scroll saw cutting is a little off. Clamp the piece to a work table, and cut it out using a scroll saw (see Figure 1-83). Use a file or rasp to remove any rough edges and even the cuts out. Finally, sand the flipper smooth. Attach a washer with a ⅜-inch hole to the bottom of the flipper using epoxy.

To design flippers in Tinkercad, go to Tinkercad and sign in or create an account. Click the "Create a new design" button. Click the "Edit grid" button in the lower-right corner, and change the units to inches, confirming by clicking the "Update grid" button. Click and drag a cylinder to the work plane from the "Basics shapes" menu on the right. Its default size is 1 inch, and that is perfect for our needs. Click to select the cylinder, and the "Shape" menu will appear. On the "Shape" menu, drag the slider for number of sides over to select the maximum of 64 sides to make the cylinder as smooth as possible. Next, drag a cylinder hole shape from the "Basic shapes" menu; hole shapes are gray with diagonal stripes. Resize the hole shape to ⅜ inch diameter, and use the "Shape" menu to increase the sides to the maximum number. Position the hole roughly in the center of the cylinder, and then click and drag to select both shapes. Select the "Align" button icon, and several circles will appear. Click the circle in the front center and on the left center to align the hole in the center of the cylinder, as shown in Figure 1-84.

Click and drag a roof shape to the work plane. Click on the shape, and then left-click and hold the rotation arrows on the right of the shape to turn it 90 degrees so that the triangle is on top. Click and drag on the rotation arrow located in the front, and turn the rectangular base side of the roof 90 degrees toward the circle, as shown in Figure 1-85. Click on the shape, and use the

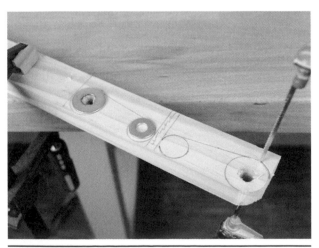

Figure 1-83   Wooden flipper.

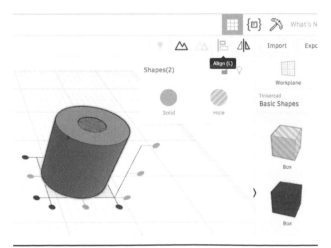

Figure 1-84   Hole alignment.

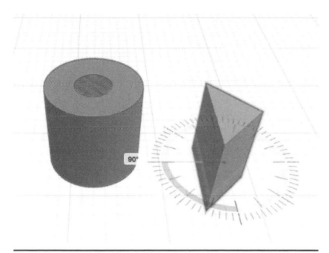

Figure 1-85 Roof rotation.

black cone to raise the bottom to the surface of the work plane. Align the base of the roof shape with the center of the cylinder, and then click on the shape and change the length to 3¾ inches. Drag a "Hole cylinder" to the work plane, and resize it to ⅜ inch in diameter. This hole will be used to slice off the pointed end of the roof shape and will be replaced with a solid cylinder. Drag the shape over toward the end of the roof so that it touches both sides. Select all the shapes, and use the alignment tool to center the cylinder. To get this shape placed as accurately as possible, change the snap grid in the lower-

left corner to ¹⁄₆₄ inch. Use the arrow keys to position the cylinder so that it touches both sides with only a small amount showing, as shown in Figure 1-86.

Click and drag to select all the shapes, and then click the "Group" icon. With the shape now split, drag a "Rectangular hole" to the work plane, and position it so that it covers the pointed end of the shape (see Figure 1-87). You may need to extend the hole to about 1½ inches so that it fully covers the shape. Drag and select both shapes, and click "Group."

To fill the void, drag a cylinder shape, and resize to a diameter of ⅛ inch. Place the shape in the space left by the hole, and then select both

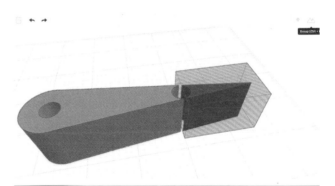

Figure 1-87 Erasing the pointed end.

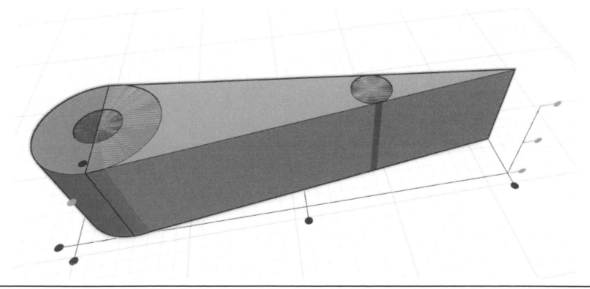

Figure 1-86 Position and alignment.

shapes and group them. Select the final shape, and drag the white box on top of it to set the height to ¾ inch. To create a hole for a set screw, drag a cylinder hole to the work plane and set it to a diameter of ⅛ inch. Rotate the cylinder 90 degrees so that it is parallel to the work plane. Use the black cone to set the height of the cylinder to ⅜ inch. Position the cylinder so that it is close to the center of the larger end and so that it goes into the main ⅜-inch hole. Use the alignment tool to center it, and then group the shapes to create the final design (see Figure 1-88). To get the file for the final design, click "Export," select everything in design, and choose the file type for your 3D printer.

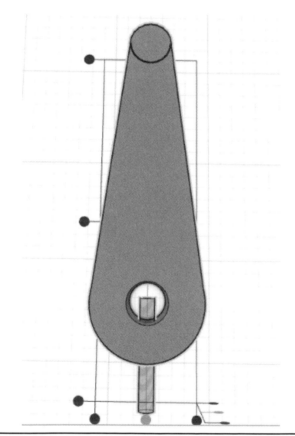

Figure 1-88 Placing the set screw hole.

Step 7: Actuators

There are hundreds of ways to build flippers and the mechanisms used to activate them. For both methods in this book, you will need two 4-inch pieces of 1- by 2-inch board. Center and drill a ⅜-inch hole ½ inch from the ends of both 4-inch pieces. Drill a set screw hole into the main hole with a ³⁄₃₂-inch drill bit, as shown in Figure 1-89. Cut and place a 2⅜-inch piece of ⅛-inch dowel through the hole. Insert and tighten a ¾-inch No. 6 screw to hold it in place. If you cannot find ⅜-inch dowel easily, most wooden hangers use ⅜-inch dowel for the pants portion of the hanger, which can easily be pried loose.

Figure 1-89 Actuator and dowel.

Step 8: Simple 1- by 2-Inch Push Stick Actuators

For the simple box, a 1- by 2-inch board and some scrap plywood can be used to construct the actuators. Start by drilling two ¹⁷⁄₆₄-inch holes in the plywood floor 6 inches away from the bottom and 4¾ inches away from the outside edge. You will need to cut a 1- by 2-inch board into two 6½ inch pieces and four 2¾-inch pieces.

Turn the pinball machine over, and measure and place a mark at 1 and 4$\frac{1}{16}$ inches from the front on both sides. Draw a line to connecting the 1- and 4$\frac{1}{16}$-inch marks. The 2$\frac{3}{4}$-inch pieces will act as guides for the push stick to travel in between. Lightly coat one side of each block with wood glue, and align the blocks with the lines and the edge of the plywood, as shown in Figure 1-90. Use 1-inch brads to secure them in place. Cut two 2$\frac{3}{4}$- by 4$\frac{1}{2}$-inch plywood rectangles. Lightly coat the tops of the 1- by 2-inch pieces you just fastened, and then nail the plywood in place over them to create the upper guide for the push rod, as shown in Figure 1-90. While the glue is drying, round off the corners of one of the 6$\frac{1}{2}$-inch pieces with a rasp or sandpaper. Sand both pieces heavily so that they will be slightly thinner than a normal 1- by 2-inch board and can slide easily in the guide. After sanding, wipe the push stick clean, and then rub paraffin or candle wax on all sides of it that will make contact with the guide. Slide the push stick in place with the curved end facing toward the center of the machine. Drill a pilot hole, and insert a ½-inch No. 8 screw partially to keep the stick from sliding out.

To make the push sticks slide back after you push them, you will need to add a hook on the actuator that is attached to the flipper. Center a small hook about ½ inch from bottom edge of the actuator, as shown in Figure 1-90. Attach a hook to the inner edge of both 1- by 2-inch pieces that make up the sides of the guide. String a rubber band between the hooks, and experiment with different thicknesses of rubber bands until the push sticks move back into place after they are pressed.

Step 9: Rod Actuators

To make rod actuators, begin by flipping the pinball machine over and drilling the two holes for the flippers in the plywood floor. Make a mark 6 inches away from the front inner edge

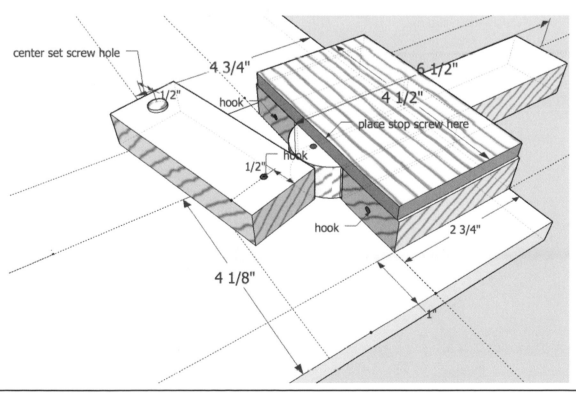

Figure 1-90 Push stick actuator.

on both the right and left edges of the plywood floor. Connect the two with a mark to make a guideline and then place a cross at 4¾ inches from the inner-left side and 4 inches from the inner-right side.

To make guides, turn a section of 1- by 2-inch board on its side so that the ¾-inch side is visible, and secure it with a clamp or vice. Center

Figure 1-91a   Drilling guide holes.

a mark ¾ inch away from the end, and drill a ¹⁷⁄₆₄-inch hole at a 90-degree angle through the board, as shown in Figure 1-91a. Remove the board from the clamp, and then cut it to a 2-inch length. Repeat the process to make a guide for the other side. Once you have cut the guides to size, you need to create two ⅛-inch pilot holes for screws to secure them to the plywood. Place the pilot holes about ½ inch from the opposite edge of the guide hole, as shown in Figure 1-91b. One of the rods will be shorter and need more space to travel, so it is necessary to cut a notch out of the guide. Orient one guide so that the screw holes are at the top, and use a small hand saw to remove a ¾-inch-square from the bottom-right corner, as shown in Figure 1-91b.

To create the push rods, cut the ⅜-inch rod to 6½ and 8½ inches. Mark a ⅜-inch drill bit at ½ inch with a tape flag, and drill out a hole in a drawer handle. These rods will take a beating, so use epoxy glue to secure them in the handle. When the handle and rods are dry, rub them down with some paraffin or candle wax and then

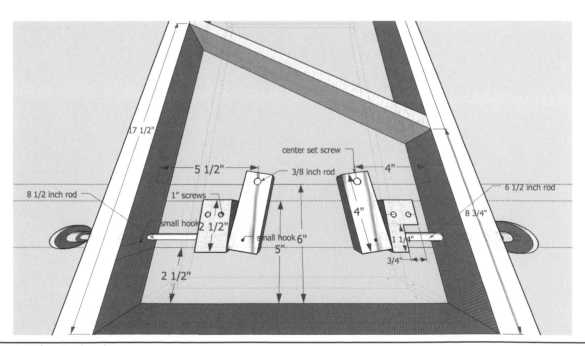

Figure 1-91b   Pushrod actuators.

place these rods into the holes on the side and then slide the guides on the ends. The guide also serves as a stop for the flippers, and its position is important. Tighten the screw on the actuator, and push the flipper onto the rod. Check the movement of the flipper when you press the rod in and note where the actuator stops when the push rod is pushed out about 2 inches. Use the set screws on the flipper or actuator to adjust the swing to your liking. When you have the movement you desire, tighten the set screw, and use 1-inch wood screws to secure the guide to the base.

To get the flippers to push back out, add a hook on both of the 4-inch actuators, as shown in Figure 1-91b. It should be centered ¾ inch on the opposite end from the rod. Place hooks on the inside wall about ¾ inch down and 2½ inches from the inside front wall. Try stringing different types of rubber bands until you get a quick return without too much tension. To keep the pushrods from flying out of the machine, place a rubber band around the rod next to the wall where the rod sits when it is at rest.

Step 10: Legs and Laptop Holder

There is a perfect playing speed and angle to achieve the perfect play on your pinball machine. We recommend that you stack some scrap blocks under the back of the machine and test it out until you find the perfect angle. Once you find the perfect angle, measure from the floor to the top of the back of your machine, and cut two 1- by 2-inch boards that length. For the machine with a ball return, you will need two 9-inch lengths, and for the basic model, they will be 8 inches. Attach them to the back with two 1¼-inch No. 10 screws, placing them ½ to 1½ inches from the sides. Before placing the legs, you might want to decide whether you want to use a small laptop as a display. In our build, we added a 15-inch 1- by 2-inch board across the top of the legs. This creates a slot, as shown in Figure 1-92,

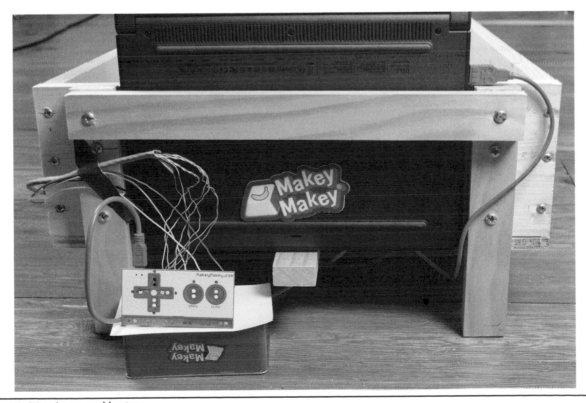

Figure 1-92 Legs and laptop.

to place a 13-inch laptop when it is fully open. You can use a small 4-inch piece of scrap wood screwed into the bottom or backside to hold the laptop at the ideal height. Before setting the height, keep in mind where you need to plug in the power supply and USB cables for the laptop. We are using an older laptop for our machine, but if you are using a newer unit, we recommend that you place a small sheet of plexiglas in front of the laptop to protect the screen. Be sure to drill holes in the plexiglas before attaching it with screws.

Step 11: Bumpers, Guides, and Obstacles

Take time to knock the ball around the machine when you get your flippers installed. Cut a few 1- by 2-inch scraps, and use them as temporary guides on the sides of the flippers. If you have access to a hole saw, try drilling a hole in the 1- by 2-inch piece and then cutting it in the center so that you have a scrap that you can fit snugly up to the flipper, as shown in Figure 1-93. Mark and cut the angles that work best for you. If you don't have a lot of scrap wood, another option is to use rubber bands and screws with nylon spacers. For our build, we used 1-inch screws inserted into ⅝-inch-tall nylon spacers to create some bumpers. You can string some wide rubber bands to create bumpers that guide the ball to the bell, as shown in Figure 1-94a.

Step 12: Wiring and Scratch Link

Since there are so many switches that need hookup wire, an old ethernet cable makes a great wiring harness. Cut the end off, measure about 10 inches from the end, and use a box cutter to splice the outer covering. Untwist the wires, and

Figure 1-93   Guides.

you find some pairs of color-coded wires. What is great about these cables is that each color pair has a striped and solid wire. Decide whether you want striped or solid to be EARTH, and you can save yourself a lot of confusion later. In the description of most of the switches, we simply told you to run a wire to EARTH or key press. You will have to define which one of the keys you want to wire it to and which sounds you want to trigger in your Scratch game. In the builds shown, the wiring harness was routed out the top of the game. For the ball return model, you will need to drill a hole for the wiring harness to exit near the top (see Figure 1-94b). You can check out the code for our game on our webpage.

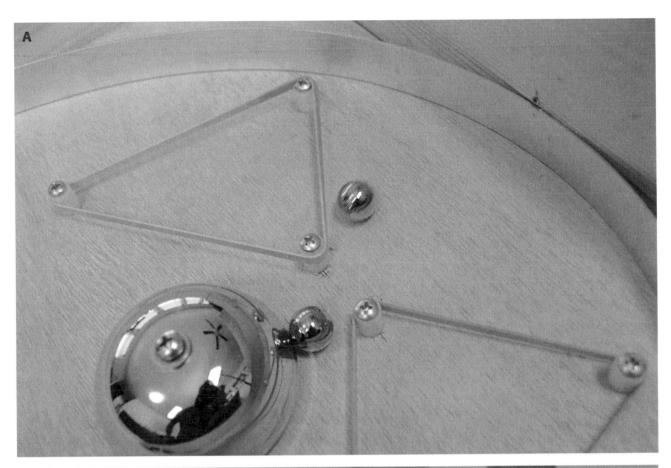

Figure 1-94 (a) Rubber band bumpers. (b) Wiring harness.

Step 13: Launcher Switch

The ball launcher is the perfect location for a switch because of the metal rod (see Figure 1-95). We will create our EARTH by drilling a small hole through the bottom corner where the rod enters the machine. Loosen the set screw on both collars, and move the spring and washer away. Strip the end of the wire, and slide it through the hole. Make a loop at the end of the wire, and then attach it to the wood with conductive tape around where the washer rests. Replace the washer, and make sure that the spring has a slight bit of tension; then tighten the set screw. We will use the collar at the end of the rod to act as contact point against the wood guide. Drill a hole in front of the bracket, and place the wire through. Secure the wire to the bracket with conductive tape where the collar contacts the wood guide. Route one wire to EARTH and the other to the key press of your choice on Makey Makey.

Step 14: Ball Return Switches

Cut a length of 1- by 2-inch wood 17¾ inches long with 60- and 120-degree angles on the ends. With the ¾-inch side up, drill three ⅛-inch pilot holes, one in the center and the others 2 inches from the end. Follow up with a ¼-inch bit to a depth of 1 inch. Flip the machine over, and place the length of wood so that the lowest edge is about 8 inches from the inner edge of the front. Secure it in place with 1-inch screws. Next, use a ruler to mark where the ball will exit the side of the machine. On the outside edge, mark down 1¾ inches, which will be the top of the hole. Use a large drill bit to drill a hole at the same angle as that of the 1- by 2-inch piece we just installed as close to the plywood as possible. With a hand saw, cut straight down to match the angle of exit (see Figure 1-96).

We recommend that before placing your holes you play your machine for a while and keep track of the spots that are somewhat difficult to hit. The holes we are creating will become "skill shots." Mark the spots, and make sure that they are above the angled 1- by 2-inch piece on the back of the board. Use a 1-inch hole saw, auger, or paddle bit to make holes at those locations.

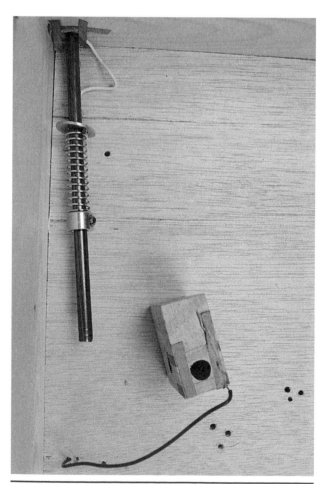

Figure 1-95 Ball launcher.

Figure 1-96 Drilling and cutting the ball exit.

With the holes drilled, use the remaining scraps of plywood to cover the back of the machine. Secure the back in place with 1-inch screws after you complete the switches for each hole. For the switches, you have several options, but one thing to keep in mind is that the ball needs to move toward the bottom of the machine after the switch is triggered. A set of simple pressure switches that are taped to the back cover and are triggered when the ball falls through the hole will suffice (see Figure 1-97a). The wires were routed to the top of the machine and covered with tape so that the ball would easily roll over them. Be sure to test your switches as in Figure 1-97b before screwing down the back cover!

Step 15: Bell Switch

If you have a metal bell and a metal ball, you might be pondering whether you have the beginnings of an awesome switch. Why couldn't you just place some conductive tape next to the bell so that when the ball hits the bell, it completes a circuit? In theory, it should work right? Actually, if you were to hold the ball there, it would work, but if you try it with a quickly moving ball, the situation is much different. Only a very small portion of the ball is making contact with the tape, and since the ball is moving very quickly, it might not trigger a connection with your Makey Makey. The Makey Makey might register this as an error in the sampling rate because the contact time and surface area just aren't big enough. To fix this problem, like all evil geniuses, we cheated. We planned where we wanted the bell, traced around it, and then positioned a roller switch just inside and behind the bell. So when the ball hits the bell, it will ding and also trigger the roller switch. This supersensitive switch is triggered by a very small amount of movement and gives Makey Makey a nice strong connection to trigger a

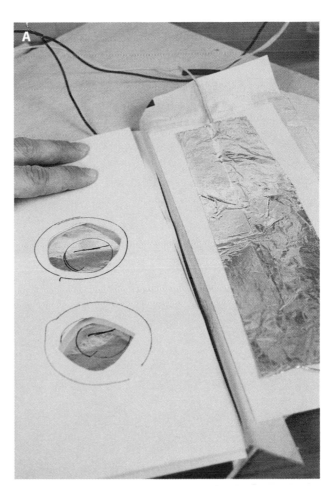

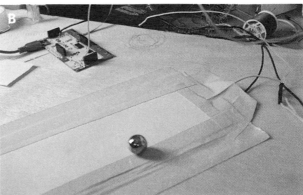

Figure 1-97 (a) Pressure switch construction. (b) Wire routing and testing.

Figure 1-98   Bell switch wiring.

key press during the sampling rate. To wire the switch, you'll connect a ground wire to pin 1 and your key press wire to pin 2 (see Figure 1-98). (You can also wire this switch so that it is always triggering a key press on your Makey Makey if you want to institute some kind of evil timer. Just saying ....)

Step 16: Rail Track Switch

You know how we just discussed how a metal ball and two conductive surfaces do not make the best switches. It's still true, however, that there are some things that you can do to make it work. One of those is slowing the ball down or extending the surface area with which it makes contact. So you will want to add some curves to your track and ensure thast the ball is making contact with both sides. You can use solid copper or steel wire to make a track (see Figure 1-99). Wire from clothes hangers works great, but you will need to lightly sand it to remove the coating. Our rail track is just under

8 inches long, so no cross ties are needed to hold it together. The start of the track needs to be in a place where the ball can be hit hard enough to roll up the rails or in a place where the ball can be guided into the rail while coming back down. We opted for the end of the curve because the ball has a good deal of speed from the ball launcher. To start, drill two ⅛-inch holes in the plywood floor about a ½ inch apart to create the starting point. Place about 1 inch of the wire through, and bend the ends over on the bottom. You will need to form the wire so that the gap is about ⅜ to ¼ inch. You can make the track twist or turn however you desire, but remember that the point is to slow the ball down so that the Makey Makey will register a complete circuit. Drill two ⅛-inch holes about ⅜ inch apart into a small scrap of 1- by 2-inch wood for the bottom of the ramp. Place a pilot hole through the scrap, and once you have it positioned where you would like, secure it with a 1-inch screw. On the backside of the machine, connect one wire to EARTH and the other to a key press on the Makey Makey.

Step 17: Swing Switch

This simple switch is made with cardstock or cardboard flag coated in foil that swings back and touches a strip of conductive tape to complete a circuit when the ball strikes it (see Figure 1-100). Start by making some small rectangles out of cardboard. Bend some wire at about 2 inches to make an L shape to form one of the posts for the swing frame. You might want to add a slight bend to the end of the wire that will become the crossbeam. This will keep the flags from sliding into the pole. Cut some cardboard flags about ¾ inch wide and 1¼ inches tall. Use heavy-duty aluminum foil, and cut a foil strip the same width as the cardboard flag. Adhere the aluminum with

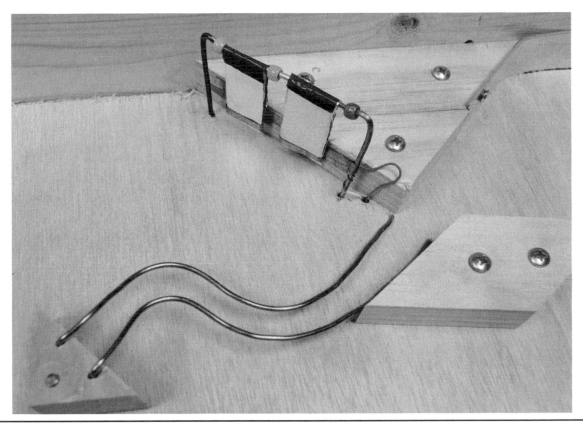

Figure 1-99   Rail track and ball switch.

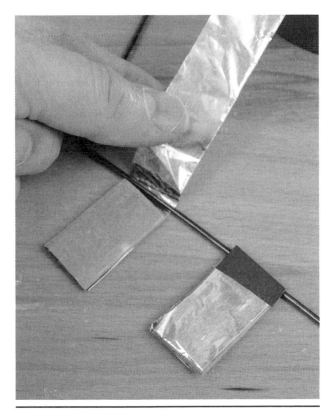

Figure 1-100   Swings with aluminum.

double-stick tape to one side of the cardboard rectangle. Place the flag next to the wire, and fold the aluminum foil loosely over it and use some double-stick tape to adhere it to the other side. Because the foil is relatively thin, use some tape to reinforce the top of the swing. We placed some beads between our flags as spacers to keep them from interfering with each other. When you have placed all the flags onto the wire, bend the leg down and snap it off at 2 inches. Position the swing about ¼ inch near a wall or obstacle. Mark the position of the legs, and drill ⅜-inch holes through the wood. Place the wire through the holes, and bend the ends over. Place a strip of tape on the wall where the swings will contact the wall. Drill a hole, and place a wire through it; then attach it to the tape. Run this wire to a key press on the Makey Makey. Connect a wire to the swing frame and route this wire to EARTH on Makey Makey.

Step 18: Basket of Nails

Vintage pinball machines often use baskets created by brads that are partially hammered to form V shapes for scoring. To create one of these, draw a V or U shape on your pinball floor. Place the brads about ⅛ inch apart, and hammer them in so that only about a half-inch of them remains above the floor. Flip the machine over, and wrap a wire around the bottom brad and bend it over. Strip a long section of wire, and wrap it around the two brads to the right and left of the center brad. Bend the rest of the brads back, and run one wire to EARTH and another wire to a key press on the Makey Makey. When the ball comes to rest at the bottom of the V shape, it will complete the circuit. It's important to note that from that point on, the circuit will

Figure 1-102   Wiring the V.

be complete, and your computer will think that your are constantly pressing that key (see Figures 1-101 and 1-102).

Step 19: Teeter-Totter Channel Switch

This switch uses a U-shaped channel made from sheet metal or posterboard that tilts on a straw. When the channel tilts, the metal channel or conductive tape touches a strip of conductive tape or fabric on the playing surface of the machine completing a circuit. Begin by making a channel out of thin cardboard or metal. If you are using sheet metal, be sure to add some tape to the sharp edges. If you use poster board later, you will coat the end with conductive tape where the channel will make contact with the playing surface. Cut a skewer ½ inch longer than the width of the channel. Use a hot-glue gun to place it almost on the center perpendicular to the channel with the skewer hanging about ¼ inch over each side. Due to the imbalance, one side will naturally rest on the playing surface; you will want that side to face the front of the

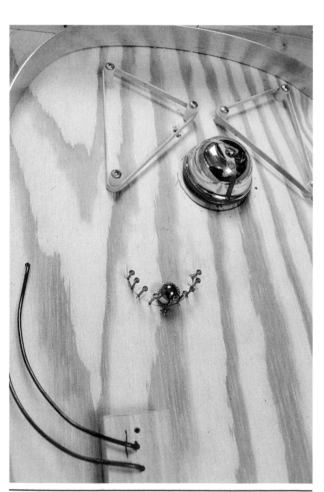

Figure 1-101   Nail basket.

machine. Speed is everything in the success of the switch, and you want the ball to be traveling at just enough speed to tilt the channel. Try moving the channel around to see where the best placement will be. When you have found the best place, mark the bottom edge and the position of the skewers. Drill a small hole just off to the side by the center of the channel and place some hookup wire through it. Strip the end of the wire, and attach it to the channel with conductive tape, as shown in Figure 1-103. If you are using poster board, run a line of tape over the wire and then to the end of the channel. Coat the end and edge with conductive tape. Put the channel back in place using the marks you made earlier, and tilt the far end down. Place some guide marks where the channel meets the playing surface when it is tilted. Just off to the side of these marks, drill a small hole and run hookup wire through it. Place conductive tape over the wire, and make a contact pad in this area (see Figure 1-104). Return the channel to position, slide a ¼-inch piece of straw over each end, and glue them in place, being careful to avoid getting glue on the skewer.

Figure 1-104 Wiring the contact pad.

Taking It Further

There are tons more possibilities for making switches on your pinball machine. Think about how you could create a Scratch game for scoring or use action figures to create a theme!

Figure 1-103 Wiring the channel.

SECTION TWO

Interactive

Makey Makey is all about interacting with the world in new and unique ways. It's also about reinventing with regular household objects. The projects in this chapter are springboards for going further with your Makey Makey. Get ready to use some new tools, output some power, and learn some creative ways to use Makey Makey as an input device! Catch your minions in the act of stealing cookies, communicate via LED, control Scratch with gestures, and more!

Cookie Jar Alarm

ARE YOUR MINIONS BEING GREEDY with their cookie calories? Set an alarm on your cookie jar with this simple Scratch trick, some conductive fabric tape, and your Makey Makey (see Figure 2-1). Your minions will be caught off guard, and you can enjoy all your evil cookies yourself!

Cost: $

Make time: 30 minutes

Skill level: 🍌

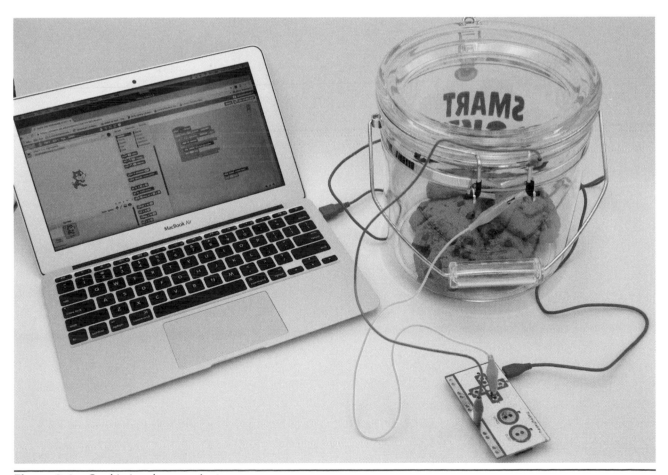

Figure 2-1 Cookie jar alarm project.

Supplies

Materials	Description	Source
Conductive fabric tape	Makey Makey Inventor Kit	Joylabz
Conductive thread	Conductive thread bobbin, 30 feet, DEV-10867	SparkFun
Cookie jar	Jar must have hinged or removable lid	Kitchen
Cookies	Something truly delicious to tempt the minions	Grocery store
Computer and access to Scratch	A free visual programming language and online community	scratch.mit.edu

Step 1: Pick a Jar

Start this project off with a clean cookie jar. Don't worry if you don't have an exact match; we also add some tips for other types of jars in this project. The two main types of cookie jars addressed in this project include one with a simple lift-off lid and another with a hinged lid that has a latch made of thick metal wire. The switch for this project will always be in the "on" position until the "cookie bandit" tries to open the cookie jar. When the switch is "off" and the circuit is broken, the alarm will trigger.

Step 2: The Latch Switch

If your cookie jar has a latch made from metal wire, you can make the latch into a switch. If your jar opens with a simple lid, skip ahead to step 3. Most latches have lower and upper components. Frequently, they are connected where the lid is hinged at the back of the cookie jar. In some cases you can simply squeeze the wire latch from the lid together and remove it from the holes in which it rests. If your lid is really firm, you may need a pair of pliers to undo the tabs on the flat metal piece that wraps around the wire. Bend the tabs out, and this should reduce the tension on the latch wire, allowing you to remove it from the holes.

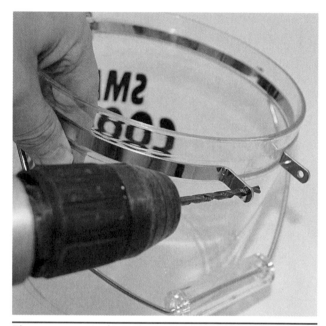

Figure 2-2 Hacking the latch.

If the holes that the latch wire rests in are already pretty tight, you may need to use a drill to drill them out a size larger, as shown in Figure 2-2. Before you drill, try cutting a slice of electrical tape in half down the middle and wrapping the end of the latch where the wire will fit into the hole, as shown in Figure 2-3. We found that wrapping it like a baseball bat handle with small overlaps worked well. Test out whether it will fit. If it doesn't fit, drill out the hole just a little larger. Place the latch wires back in the hole, as shown in Figure 2-4, and use an alligator clip from EARTH and attach it to the metal of the bottom part of your latch.

Attach another from the space key on the Makey Makey to the top of your latch, as shown in Figure 2-5. Make sure that no current travels between the two latch pieces when the latch is not connected. When you close the latch, you should see a LED light up on the Makey Makey signaling a complete circuit.

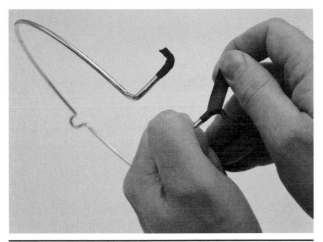

Figure 2-3 Wrapping the latch with tape.

Figure 2-4 The latch taped and in place.

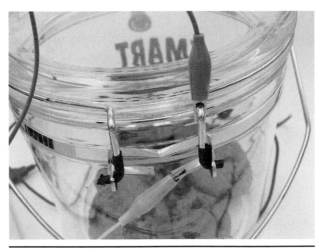

Figure 2-5 Alligator clips in place.

Step 3: Lift Lid Switch

If your jar has a hinged or simple lift-off lid with no latch, as in Figure 2-6, making a switch is just as easy. Place the lid on the jar and observe which surfaces make contact. Attach conductive fabric tape or copper tape around the rim of the lid where it meets the top of the jar, as shown in Figure 2-7. Attach a 2-foot piece of conductive thread or a very thin piece of wire to the conductive tape on the top of the lid and under the tape. This will allow for the lid to be removed freely. Clip an alligator clip to the conductive thread and the other end to the SPACE input on your Makey Makey. On the main part of your cookie jar, run a piece of conductive tape from inside the jar to the outside base, as shown in Figure 2-8. Clip an alligator clip to the tape and the other end to an EARTH input on your Makey Makey. When you place the lid on the cookie jar, a connection should be made (see Figure 2-9).

Figure 2-6 Lift-off lid.

Figure 2-7 Conductive tape.

Figure 2-8   Tape and alligator clip on base.

Figure 2-9   Connection made!

Step 4: Simple Scratch Alarm

You are going to incorporate three different logic points in Scratch to make a simple looping alarm. Start with the "when flag clicked" block as before, but instead of a "when key pressed" block, navigate to the "Sensing" palette for the "key space pressed block." By nesting this block inside the if of an if/else statement, you can make an alarm sound until the space key is pressed (see Figure 2-10). This code will stop the alarm as long as the cookie jar lid is closed. Place your "play sound" block inside the else of the statement and once the jar is opened, the alarm will sound. However, the alarm will stop after you press the space key, and in order for your alarm to be more effective, you'll want to make sure it is constantly attempting to sound the alarm. To do this, wrap a "repeat until" block around your if/else statement. You'll need to place a different key press here, or the alarm will be turned off again after pressing the space key. See the full program in Figure 2-10.

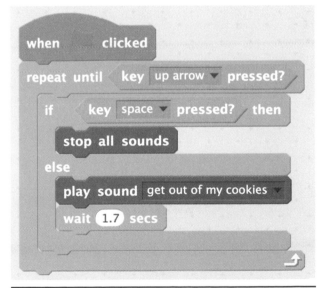

Figure 2-10   Sound the alarm in Scratch.

Step 5: Keep Minions from Eating Cookies

You can hide your computer in a kitchen cabinet, as we did in Figure 2-11a, or create this same program on a Raspberry Pi and hide it in your kitchen accordingly. Now, when your minions attempt to steal cookies, an alarm will sound as the cookie jar is opened (see Figure 2-11b)!

Taking It Further

What if there was a way to turn off the alarm? Maybe your minions should have to complete a lot of exercises so that they can turn off a fancy exercise combination lock and then be rewarded with a cookie? Try incorporating the lock-box idea from Project 15 with this cookie jar, and take your cookie jar alarm to the next level.

Figure 2-11   (A) Hide your computer in a cabinet.

Figure 2-11   (B) Minion burglar thwarted!

Makey Makey Light-Up Morse Code Machine

THERE ARE TIMES WHEN AN EVIL GENIUS needs to communicate with a secret code. Unfortunately, our minions don't have the best hearing, but they are very enraptured by blinking lights. Build this LED Morse code tower to

communicate with your minions by blinking lights (see Figure 2-12).

Cost: $

Make time: 30 minutes

Skill level: 🍌🍌

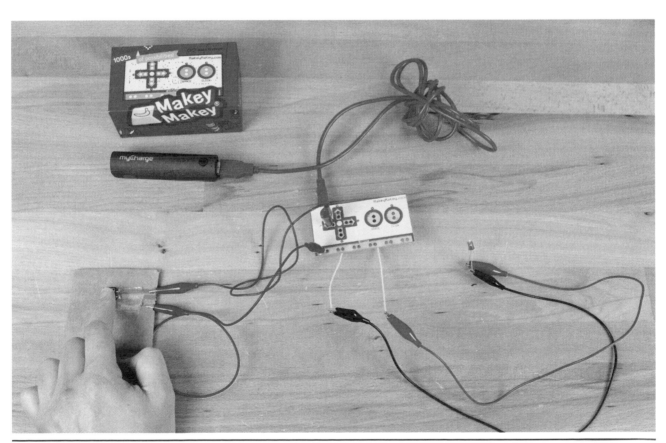

Figure 2-12   Makey Makey Morse code wiring.

Supplies

Materials	Description	Source
3D printer	Optional	School or public library
Light-emitting diode (LED)	—	SparkFun or Amazon
10-kΩ resistor	Pack of resistors 10-kΩ COM-11508	SparkFun
Push button	Momentary pushbutton switch, 12-mm COM-09190 ROHS	SparkFun
Conductive tape	Makey Makey Inventor Kit	Joylabz
Scissors or knife	Box cutter or craft knife	Hobby store
Computer and access to Tinkercad	A simple online 3D design and printing app	tinkercad.com

Step 1: Make a Tower

If you have access to a 3D printer, you can easily design a small tower to house your Morse code machine using the free online design tool Tinkercad, available at Tinkercad.com. You can sign up for an account or use an existing provider such as Google to sign in. Once you have signed in, click the "Create new design" button.

Start by clicking the "Edit grid" button located in the lower-right corner of the work plane. Change the units to inches, and click "Update grid." Click and drag a cylinder from the "Basic shapes" menu to the work plane. Change the diameter to 2 inches by selecting the shape and then entering the dimensions or dragging the white boxes located on the corners. Use the white box at the top center of the shape to adjust the height to 2 inches. With the shape selected, you will notice that a menu of characteristics appears in the upper-right corner of the work plane. To produce a smoother cylinder, maximize the number of sides by setting the slider to "64," as shown in Figure 2-13. You will need to do this to all the cylinder solid and hole shapes you create for this project. Drag a hole-shaped cylinder to the work plane. Adjust the diameter and height to $1\frac{7}{8}$ inches, and increase the number of sides to 64.

Click and drag the hole shape, and try to center it in the solid cylinder. Don't worry if you're off a bit; we will fix that later with the "Align" tool. Drag another cylinder-shaped hole to the work plane, and adjust the diameter to 1 inch and the height to $2\frac{1}{2}$ inches. Also remember to adjust the number of sides to 64. Center this hole in the solid cylinder as well. Click and drag just above the shape on the work plane, and select all the shapes. Click on the "Align" tool located in the upper-right corner of the work plane, as shown in Figure 2-14. To center the hole shapes, click on the alignment circle that appears in the front center of the cylinder and the alignment circle that appears to the left center of the cylinder. We need to make a hole in the bottom of the cylinder tower to provide a passageway for the alligator clips. Drag a simple box shape to the work plane, and adjust it to ½ inch on all sides. Drag the box into the cylinder so that it is visible from the inside and outside. Click and drag the mouse, and click on the "Group" tool that appears to the left of the alignment tool.

The top section of the tower is removable so that changing the LEDs and attaching alligator clips are easy and hassle-free. Drag a solid cylinder shape to the work plane, adjust the diameter to $1\frac{1}{8}$ inches, and increase the sides to 64. To create a hollow area in the shape, drag a cylinder-shaped hole to the work plane, and

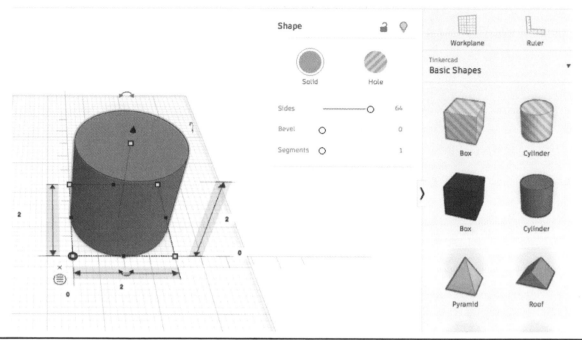

Figure 2-13 Setting the slider.

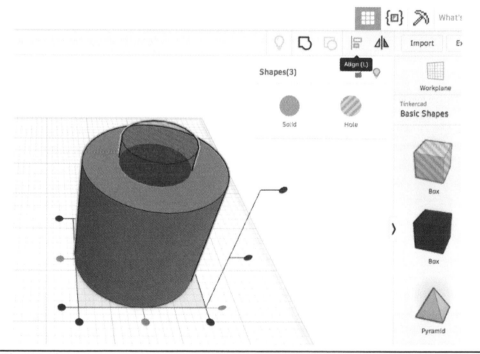

Figure 2-14 "Align" tool.

set the diameter to 1 inch and the height to ⅞ inch. Remember to adjust for the maximum number of sides. To create a hole for the led at the top of the tower, drag a hole shape to the work plane, adjust the diameter to ⅛ inch, and increase the sides to 64. Roughly center the two hole shapes in the smaller cylinder, and then use the alignment to perfectly center them, as shown in Figure 2-15. Click on "Group," and you now have the top portion complete.

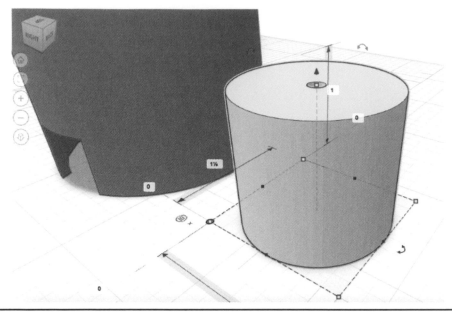

Figure 2-15   Perfect center.

To complete the tower, let's create a tube shape at the top of the base for the top to slide into. Drag a tube shape onto the work plane, and then adjust the height to ¼ inch and the outer diameter to 1½ inches. You will notice that when you select this shape, you not only can increase the maximum number of sides to 64, but you can also change the wall thickness. Enter 0.12 for this value. Depending on the type of filament and shrinkage that occurs with your 3D printer, you may need to adjust this value if the top is too snug. Put the tube in place by selecting the shape and then dragging the black cone that appears at the top to a height of 2 inches. Center it over the base, and then click and drag to select both shapes. Use the alignment tool to perfectly center it, and then click "Group" to make it a solid shape, as shown in Figure 2-16. You also might consider flipping the smaller upper portion to get a better print. Press the "Export" button on the right, and choose the file type needed for your printer.

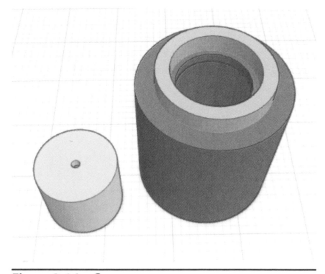

Figure 2-16   Group.

One-Banana Hack: If You Don't Have Access to a 3D Printer

If you don't have a 3D printer, you can easily use a plastic bottle to make a tower. You can also use a nail if you don't have a drill. Use a ⅛-inch drill bit to drill a hole in a plastic bottle top. You will need a box cutter to cut a hole near the bottom of the bottle so that the alligator clips pass through the bottle. Place the LED through the hole, and then clip the wires to the legs. Carefully tighten the lid and connect the wires as shown in Figure 2-17.

Step 2: Assembly

Slide the LED legs through the ⅛-inch hole in the top portion of the tower. Separate them so that each leg touches the sides. Polarity matters on a LED, so remember that the long leg of the LED should be the positive leg. Clip an alligator clip onto the positive leg (we suggest using a red alligator clip to remember that it is positive), and clip a black or gray alligator clip to the negative leg. Ensure that the protective covering is slid down over the end of each clip so that the metal from the alligator heads do not touch. Run the wires into the hole on the base and out the bottom, as shown in Figure 2-17. Attach the positive leg of the LED to the KEY OUT output pin on your Makey Makey and the negative leg to EARTH.

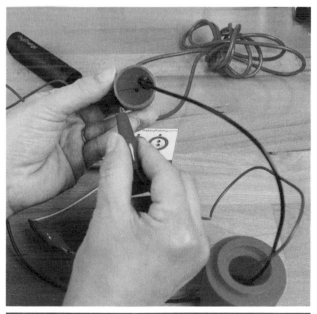

Figure 2-17 Alligator clip to LED legs minding polarity.

Step 3: Wire a Button to Cardboard

For our key press, we will use alligator clips to wire a commonly used electronic component, the button! You do not have to wire a 10-kΩ resistor to your button, but if you eventually want to program it as a digital pin with an Arduino, this can be good practice. So why not get some practice wiring components this way? You'll need three traces of copper tape on your cardboard square. Bend the legs of your button so that the metal can rest along the copper tape and the ends can still poke down into the cardboard. You only need the front feet of the button to touch each copper trace. The back legs of the button can be pushed into the cardboard for stability. Use a 10-kΩ resistor between what you are using as the negative side of you button, and the other leg of the resistor will be placed on the copper tape. Bend the legs of the resistor by bending them like a zigzag so that the legs will have more connection to the copper tape. Use a regular piece of clear tape to secure the resistor on top of the copper tape traces (see Figure 2-18). You'll clip the key press input to the positive pin of your button, and you'll connect EARTH to the tape trace connected only to the 10-kΩ resistor (see Figure 2-18). Make

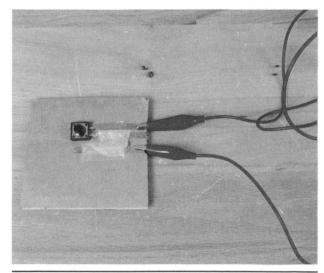

Figure 2-18 Wiring a button to cardboard.

sure that you plug your Makey Makey into the computer, and you are ready to start sending secret messages! (You can actually just press the copper tape traces to turn your LED off and on, but where is the fun in that? Now you know how to wire up a button should you need to, you can actually wire a signal to the middle tape trace, but with this cardboard button, you also get a great clicking sound reminiscent of a real telegraph machine!

Step 4: Send Secret Messages

Print out a copy of the Morse code alphabet, and teach your minions the long and short of it. Practice making long and short button presses and creating words with the "blink" of a LED (see Figure 2-19)!

Taking It Further

Is one LED enough? How could you add more LEDs to this design or your own? What other ways could you use a LED to interact with a program in Scratch? If you liked 3D design, try creating a cover for the LED out of translucent filament for light play. Or do some research and discover a way to make the light brighter!

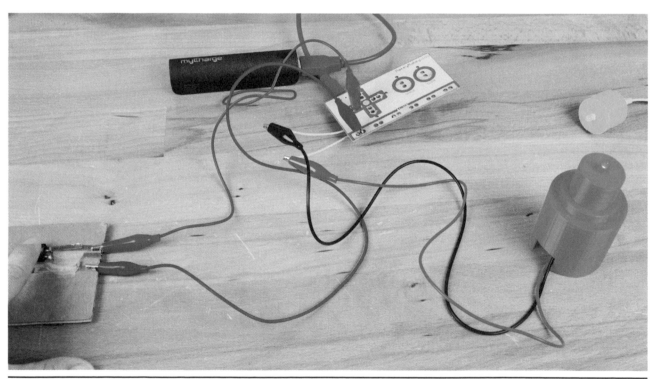

Figure 2-19 Send messages.

Makey Makey
Etch-a-Processing Sketch

Remember how much the minions used to love playing with your Etch-a-Sketch? What if you could create a new one with a simple Processing program and a Makey Makey? Find a red box, some milk lids, and let's get started. If you haven't used Processing before, it is a really cool programming language for making interactive digital art. To create our own Etch-a-Processing

Sketch, we only have to write a few lines of code that will track our mouse movement (see Figure 2-20).

Cost: $

Make time: 30 minutes

Skill level: 🍌🍌

Figure 2-20 Makey Makey Etch-a-Processing Sketch

Supplies

Materials	Description	Source
Bottle lids	Screw-on plastic bottle lid about 1½ inches in diameter	Recycling
Cardboard box	Box size about 10 by 6 by 2 inches	Recycling
Bare Conductive paint	10-mL tube	www.bareconductive.com
Screws, nuts, and washers	Two 1½-inch $^{8}/_{32}$-inch Phillips pan head screws, four 1-inch $^{8}/_{32}$-inch Phillips pan head screws, six $^{8}/_{32}$-inch machine screw nuts, ten No. 8 flat washers, two ½-inch No. 4 pan head screws	Hardware store
Nylon spacer	Two ½-inch-tall nylon spacers with an OD of $^{3}/_{8}$ inch and an ID of 0.171 inch	Hardware store
Computer with Processing program	A flexible software sketchbook and programming language for beginning coders	https://processing.org/

Write a Sketch in Processing

Step 1: Look Around Processing

Download the Processing IDE from https://processing.org/. Create a new sketch, and let's take a moment to look around at the Processing environment. It's pretty straightforward. You have your sketch window, a "Play" button that will run your code (which will run your sketch in a new window), and a "Stop" button. If you are new to this type of programming, I've included comments for every line to explain how the sketch works. You do not need to include the comments in your own sketch. However, commenting also helps you to debug when things go wrong. "//" is used to note a comment, so that a computer will not read the comment as a line of code. Instead, the computer will skip anything after "//" and read the next line.

Step 2: void setup()

Everything you write in `void setup()` will run only once when you start the sketch. Inside the parentheses we will set the size of our screen area and then tinker with color areas so that you can see how to change the sketch if you get another idea you want to try.

Let's set the size of the screen area and background color type. The `setup` function will run everything inside the brackets once at the start of the sketch. Remember to include semicolons at the end of each line of code.

```
void setup()
{
size (500,500);
background (100);
}
```

Now click the "Play" button to run your sketch. A gray square should appear. Now, to make our Etch-a-Processing Sketch window, change the size and add a frame rate so that the frame rate will refresh when you are drawing, allowing you to track mouse movement from frame to frame. To resemble an Etch-a-Sketch, we want to set the background to white by setting the color code to 255.

```
void setup()
{
 size (1000,800);
 frameRate (15);
 background (255);
}
```

Step 3: void draw()

To track mouse movements, we only need four more lines of code. The `void draw()` function will continually run the lines of code between the brackets until the sketch is stopped. You must use this function to track the movement of the mouse while the sketch is open. You will set your stroke by typing "stroke" and include the color for your stroke in the (). For tinkering purposes, let's try "5" inside the parenthesis. If you run the sketch like this, nothing will happen because we haven't tracked our mouse movement yet.

```
void draw()
{
  stroke (5);
}
```

Step 4: Track Mouse Movement

We not only want to track our mouse movement, but we also want to draw a line between the mouse's previous *x,y* location and the current *x,y* location so that our sketch will resemble the function of an Etch-a-Sketch. To do this,

we will use `line()` because it allows us to draw a line between two paths. Inside the parentheses (`mouseX, mouseY`) will find the current location of the mouse, and (`pmouseX, pmouseY`) will draw a line from the current location to the previous location of the mouse from the previous frame.

```
void draw()
{
  stroke (5);
  line (mouseX,mouseY,pmouseX,pmouseY);
}
```

Try changing the stroke color to the number 240. What happens? We want to make our line look more like the gray color of an Etch-a-Sketch, so we will set our stroke color to 120.

Step 5: Test

Your Processing sketch should now resemble Figure 2-21 without the comments. It's time to test your sketch and begin drawing with your mouse in the "Sketch" window (see Figure 2-22).

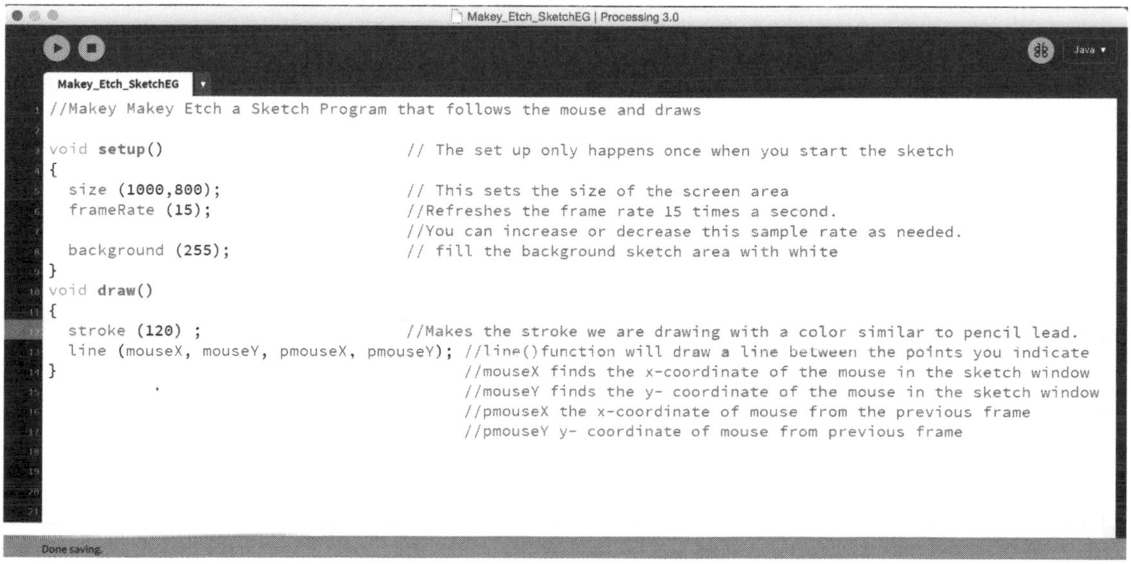

Figure 2-21   Processing sketch with comments.

Figure 2-22 Testing mouse drawing in "Sketch" window.

Step 6: Tinker with strokeWeight()

You might be wondering how to change the size of your drawing line. The `stroke` function defaults to a width of 1 pixel, and if you want to change it, you'll have to add `strokeWeight()`. Try adding `strokeWeight (100)` to your sketch. Once you are happy with the size of your drawing pencil, it's time to make your box and hook up your Makey Makey.

```
void draw()
{
  stroke (5);
  strokeWeight(100);
  line (mouseX,mouseY,pmouseX,pmouseY);
}
```

Make an Etch-a-Sketch Box

Step 1: Prepare Box and Drill Holes

With the lids on the box, decide the best placement. For this build, we centered the lids 2 inches from the side and 1½ inches from the bottom. Mark the bottom and side of the lid, and align it with those marks. Use a pencil to lightly trace around the lid, as shown in Figure 2-23. Measure up from the bottom of the box to where the center of the lid is located, and place a mark at the center and at the same distance from the bottom on both sides. It's a good idea to draw a line across the center.

The lids will become the dials used to control our sketch machine, but we don't want them to rotate all the way around because that would twist the wires too much. To achieve this, we are going to place some stop screws just below the center that will also act as connections for our directional controls. Measure in about ⅛ inch from the side, just below the center, and turn the screw over so that the head is resting against the box. Trace around the head as shown in Figure 2-24, and then mark the center. Repeat this step for the other side. Use a drill, craft knife, or pencil to make a hole for the screw. Also place a hole in the center of the circle for the screw on which the dial will rotate.

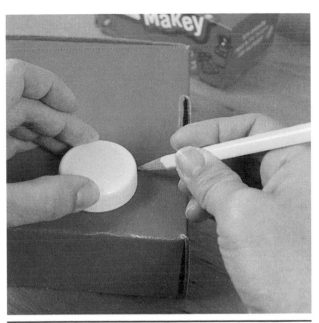

Figure 2-23   Trace around the lids.

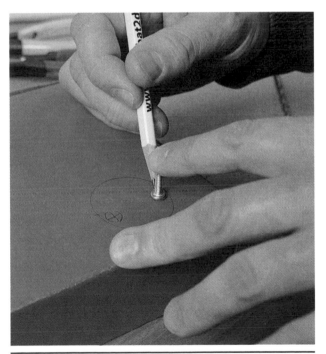

Figure 2-24   Determine screw placement.

Step 2: Prepare the Box with Conductive Paint

Use the Bare Conductive paint to make larger contacts areas for the dials. Be sure to paint the area around where the screw will poke through the cardboard so that it will make contact. Leave at least a ½-inch space in the middle so that when the dial is centered, it will not make contact with either connection and will act as a stopping point for sketching (see Figure 2-25). Place the box aside and allow the paint to fully dry.

Once the paint is dry, slide a flat washer onto a 1-inch screw, and place it into the holes you drilled for stop screws. You can see the stop screws and flat washers in Figure 2-26. Notice that the flat washer makes contact with the Bare Conductive paint. On the opposite side place another washer, and then fasten a nut firmly to the screw. Repeat this step for the other three holes.

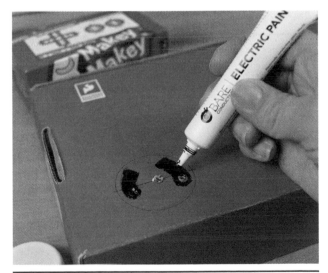

Figure 2-25   Bare Conductive paint pen.

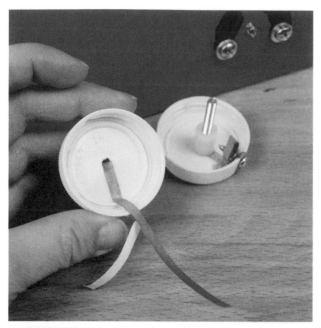

Figure 2-26   Screw placement and preparation of controller lids.

Step 3: Transform Lids into Controllers

Use a clamp or a pair of pliers to hold a plastic lid, and drill a $\frac{5}{32}$-inch hole in the center of the lid. This hole will connect the controller to the box and also function as the EARTH input. Now turn the lid on its side and drill a $\frac{2}{32}$-inch hole near its bottom edge. Later we will insert a set screw here to constrict the movement of the controller.

On the inside of the lid, run a piece of conductive fabric tape from the hole in the center across the lid, up the side, and over the hole that you drilled for the stop screw. Run the tape up the outside wall and back to the hole in the center of the top, as shown in Figure 2-26. Place a $1\frac{1}{2}$-inch pan head screw into the hole, and place a spacer on the bottom side. Turn a $\frac{1}{2}$-inch No. 4 screw into the hole on the side, piercing through the tape (see Figure 2-27). Be careful not to twist the tape around the screw. You can ensure that the screw goes through the tape by holding the tape with a fingernail as you turn the screw through the tape. Cut a $1\frac{1}{2}$-inch piece of conductive tape, wrap it around the No. 4 set screw, and then extend it about $\frac{1}{4}$ inch over the edge of the lid. Fold the tape over, and stick it to itself back down to the set screw to create a flexible brush, as shown in Figure 2-27. This will extend your EARTH connection and brush across the conductive paint inputs on the box. Before securing the controllers to the box, insert the screw into the center hole you drilled and observe whether the brush is making a connection with the conductive paint, as shown in Figure 2-28. If it does not, you may need to make the conductive tape brush longer. Also check to see that the brush does not touch either side of the conductive paint while the dial is centered. If so, shorten the length of the tape brush or you will not have very good sketching control.

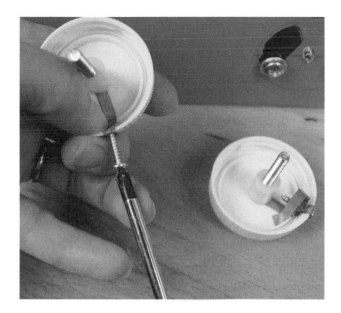

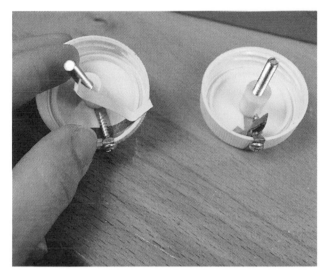

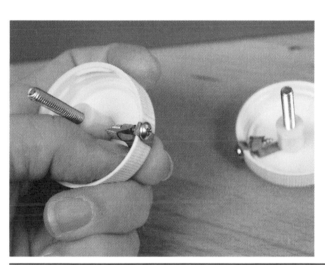

Figure 2-27   Lids as EARTH brush.

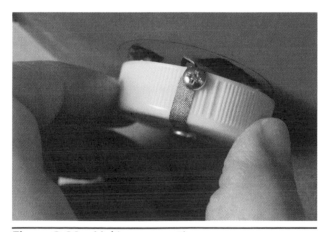

Figure 2-28   Making connections.

Step 4: Attach the Lids to the Box

Once you have made any necessary adjustments, place the controller back through the hole, and center your conductive tape brush so that it is not touching either conductive paint input. On the inside of the box, slip on a washer and fasten a nut on the screw, but leave the screw so that it is still turnable. We are going to lock the position of the nut in place with a second nut. This will allow the dial to turn but not come loose. With a pair of needle-nose pliers, hold the first nut

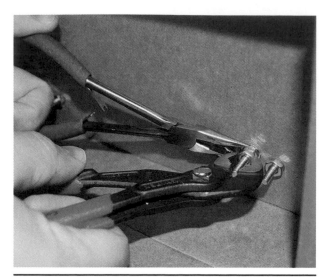

Figure 2-29   Attaching the lids from the inside.

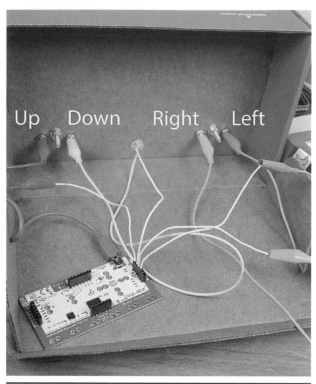

Figure 2-30   Hooking up your Makey Makey.

firmly in place, and use a second pair of pliers to tighten the second nut as much as possible, as shown in Figure 2-29. You should be able to move the dial left to right, noting that when the set screw touches the stop screw, the dial's rotation is limited.

Step 5: Hook Up Your Makey Makey and *Play*!

Now you are ready to hook up your Makey Makey and get to sketching! On the back of the Makey Makey on the right side you'll see four pins for mouse movement. Place a white jumper wire in each pin. Place an alligator clip on each jumper wire, and then connect your alligator clips to each screw, as shown in Figure 2-30. (Remember that we are wiring from the backside, so the directions may seem confusing at first.) Make sure that none of your wires are touching. You can tape the alligator clips to the inside of the box if you wish. Clip an alligator clip to the center screw and an EARTH input for each controller. Now you are ready to close your Makey Makey inside the box. Plug your

Makey Makey into the computer, and press "Play/run" on your Processing sketch. Turn the dials, and etch your Processing sketch! If you notice that your mouse is continually dragging, double-check to make sure that none of your alligator clips are touching and that you have enough space between conductive paint inputs on your box. While you play, your tape brush may loosen, so check to make sure that it isn't continually making contact as well.

Taking It Further

You can install Processing on a Raspberry Pi and make this Etch-a-Sketch practically portable! Add a PiTFT (touchscreen or not) to the box to have one fully enclosed, transportable project.

Makey Makey Musical Hoodie (Interactive Clothing)

COULD IT BE POSSIBLE TO MAKE something that is both warm and supercool at the same time? In this project, you will learn how to use conductive fabric and thread to make your own clothing interactive using Makey Makey, a speaker, and Raspberry Pi. This project will give you the skills to make a hoodie to play your own evil theme song and remix it on the move (see Figure 2-31).

Cost: $

Make time: 30 minutes

Skill level: 🍌 🍌 🍌 🍌

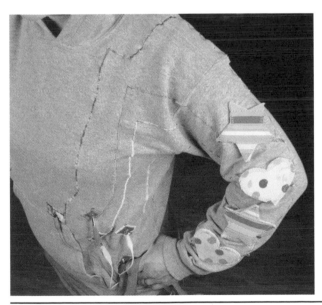

Figure 2-31   Musical hoodie.

Supplies

Materials	Description	Source
Conductive thread	Conductive thread bobbin, 30 feet, DEV-10867	SparkFun or Amazon
Conductive fabric tape and conductive fabric	Supplies from Makey Makey Inventor Booster Kit	Joy Labz
Raspberry Pi 3, Model B		
HDMI cable or ethernet cable, USB keyboard, and USB mouse	—	Radio Shack or electronics store
Portable rechargeable speaker	Must have ⅛-inch jack	
Raspberry Pi 3 and access to Scratch	A free visual programming language and online community	scratch.mit.edu
Hoodie	Any item of clothing you'd like to hack	Your closet
Fabric	Assorted fabric materials for your fabric switch	Joann's
Ribbon	Assorted ribbons for creating soft leads	Joann's
Thin-gauge hookup wire	50 feet of 30 AWG wrapping wire	Radio Shack (2780502)

Make a Fabric Switch

Step 1: Prepare Fabric Iron Fusible Interfacing to Fabric

You need to make your fabric switches a little stiff so that the switch does not accidentally complete the circuit with natural movement while you are wearing your interactive hoodie. To do this, use a semistiff fusible interfacing and adhere it to the fabric of your choice. Follow the directions on your interfacing to fuse it to fabric. Depending on the shape of your fabric switch, you will need about a 16- by 12-inch block of fused fabric for making shape cutouts. Cut your interfacing about ½ inch smaller than your fabric. Most interfacing requires a damp cloth between a hot iron and the interfacing, which will help to fuse it to the fabric.

Step 2: Create a Template and Cut Out Shapes

Once you've prepared a block of fabric, create a template or use a precut foam shape to trace

your desired shapes on the fused side of the fabric. You need to make three identical shapes for each switch. The bottom layer will be the positive contact (the key press), the middle layer will insulate your circuits, and the top layer will function as the EARTH contact. You'll be creating four switches, so you'll need 12 layers in total.

Step 3: Connect Conductive Switch Pads

Following the template in Figure 2-32, use the conductive fabric tape to attach a small square of conductive fabric to the fused side of the bottom layer, and repeat for the top layer. Use a second piece of tape to create a conductive tape tab by folding the tape over on itself, as shown in Figures 2-32 and 2-33. For the center layer of fabric, you will cut a hole as in Figure 2-33. This hole will allow the fabric switch to complete the circuit only when someone presses on the center of the fabric switch. Place the center layer on top of the bottom layer, and make sure that you do not see any conductive fabric around the edges,

Bottom (Key Press)   Top (Earth)

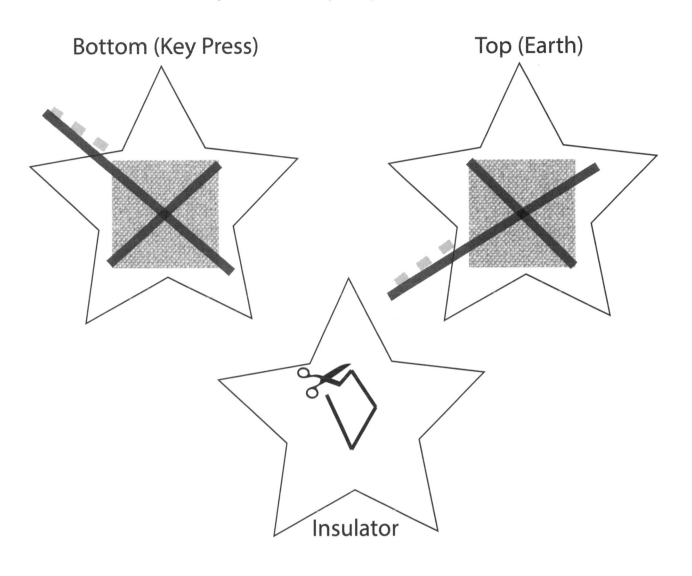

Insulator

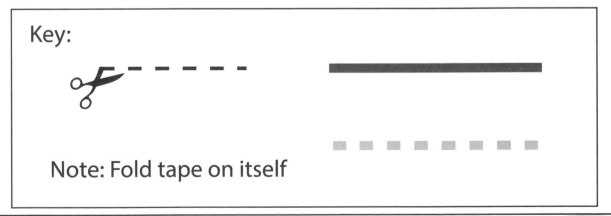

Figure 2-32   Fabric switch template.

Figure 2-33 Fabric switch construction.

as in Figure 2-34. You should only see the fabric through the hole you cut. Flip the middle layer to the top, and make sure that you do not see any conductive fabric or tape because they will short circuit your switch from this view as well. Place all three layers together for the next step.

Step 4: Sew the Switch Together

Using the zigzag stitch on a sewing machine, make sure that you adjust the width, and sew together your fabric switch as in Figure 2-35. If any fusing is visible on the outside of your stitch, you can trim away excess fabric. Just make sure that you do not trim the conductive tape lead away!

Figure 2-35 Sew the switch together.

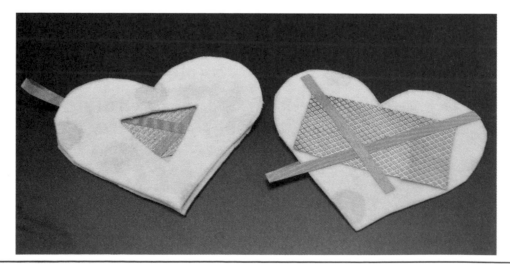

Figure 2-34 Check layers for short circuits.

Step 5: Repeat for All Shapes

Repeat steps 3 and 4 for all your fabric switches.

Step 6: Test Your Fabric Switches

Now that you have four fabric switches, let's hook up your Makey Makey to each fabric switch pad and test them to make sure that the circuitry is effective before sewing circuits onto your hoodie. Clip an alligator clip to an EARTH input and one conductive tape pad. Then clip a second alligator clip to any key press and the other conductive tape pad, as shown in Figure 2-36. Make sure that your Makey Makey is plugged into the computer, and press on the fabric to see if your switch will make a good connection. You may also want to see if your switch is too sensitive. To test this, grab your

switch while it is connected to the Makey Makey with both hands. Emulate movement by wiggling it back and forth. If your switch activates with slight movement, you'll want to add another layer of fused fabric between your positive and negative fabric layers.

Sew Circuitry

Step 1: Adhere Shapes

Use stretchy fabric glue to adhere your shapes to the sleeve of your hoodie. Or you can hand stitch the switch with nonconductive thread. If you use this method, be careful not to apply too much pressure to your switch or it will remain activated, and you'd rather debug now than after sewing all of your circuitry.

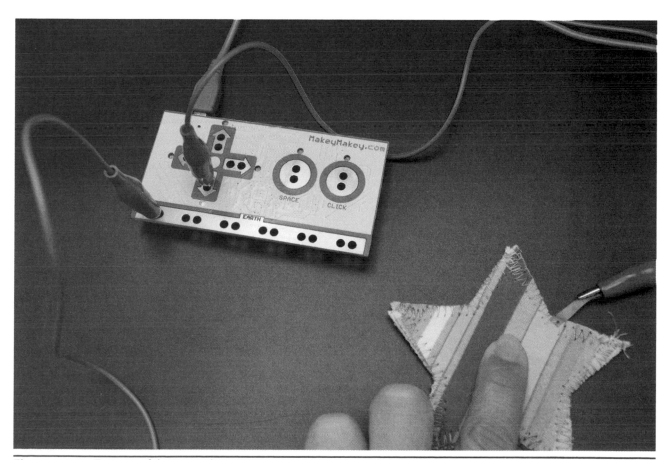

Figure 2-36 Test your fabric switches.

Step 2: Make Conductive Patches

If you are using the fabric square from the Makey Makey Inventor's Booster Kit, use an iron to adhere fusible interfacing to the back of the fabric. This will keep the ends from fraying and give it a little more strength.

Now cut four to five small circles or squares of conductive fabric, and use the fabric glue to adhere them to your hoodie (interfacing side down). Make sure that you have a small square of conductive fabric for each switch (we are using four fabric switches in this project) plus one conductive patch for EARTH that will ground all fabric switches. You may also want to sew around the conductive fabric with a straight or zigzag stitch to ensure that the edges won't fray after washing.

Step 3: Map Circuitry

Using a water-soluble pen or fabric marker, map out the circuitry from the conductive patches to each fabric switch (see the template in Figures 2-37 and 2-38). You will have a circuit trace from one side of each fabric switch to a conductive patch. Remember that each fabric switch will function as a different key press. This project is actually a good beginner project for getting hands-on experience with sewing circuitry for Arduino wearable projects. Instead of components, you'll have practice sewing your own handmade switches to each conductive patch that you will then wire to a pin on the Makey Makey. For your EARTH connection, you will actually ground all the outer-edge tape leads from your fabric switches with one continuous conductive thread trace to the EARTH conductive patch. (In typical wearable sewing, this would be your grounding trace.)

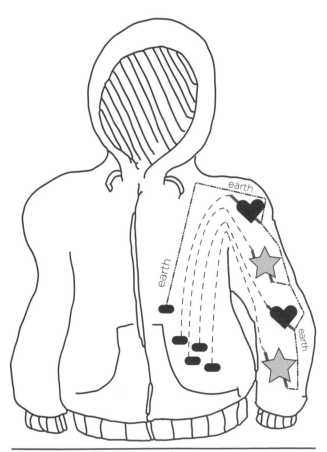

Figure 2-37   Circuitry template.

Figure 2-38   Map conductive traces.

(If you decide that you like sewing circuits, this project is actually very similar to sewing wearables with a Lilypad Arduino microcontroller. You would still sew your grounding circuit all to the negative pin. The main difference is that you would sew each positive lead to a pin on the Lilypad rather than to these conductive patches.)

Tips for Working with Conductive Thread

- Because of the steel laced inside it, conductive thread is sticky compared with regular thread. It tends to knot and stick on itself quite frequently when sewing. If your thread seems to knot up, allow your sewing needle to hang, and let the thread unwind itself. Gently pull at knots to straighten.

- Conductive thread actually tears quite easily, so be careful when pulling on it so that you don't break the thread apart. When tying knots at the end of a circuit trace, you can tie the knot and pull the thread to break it instead of cutting it.

- Make small and even stitches when hand stitching rather than long, loose stitches. Long stitches tend to be too loose and might accidentally short your circuitry.

- Take the time to check the underside of your work as you sew. Because this thread tends to stick to itself, it's good to make sure it isn't knotting up on the inside of the hoodie.

Step 4: Sew Circuit Key Presses

You have two options for sewing your circuitry. The first requires more control over machine sewing, and the second requires more time but less experience with a sewing machine. Choose wisely, oh evil one.

Option 1: Appliqué/Couching Technique

For the appliqué method, you will not actually be sewing with your conductive thread. Instead, you'll be holding your conductive thread along the circuitry path you drew in step 3 and then using an appliqué stitch (with nonconductive thread) to cover and insulate the conductive thread trace. A couching technique is a common sewing technique where you hold one type of thread, cord, etc. against the fabric and cover that thread, cord, etc. with a stitch to hold it in place. One of the great things about using this technique on your hoodie is that you will automatically insulate your circuit traces.

To sew an appliqué stitch, it is best to follow the specific instructions on your machine. Generally, you have to tinker with the tension and stitch-width settings on your machine to achieve the look you want. Practice on scrap fabric before sewing on your hoodie, and make sure that you always use a tear-away stabilizer on the underside of the fabric when sewing an appliqué stitch.

For extra security in creating a good connection, use conductive fabric tape to connect the conductive thread to your conductive patch, as shown in Figure 2-39. Cover all four sides of your patch with fabric tape if you didn't

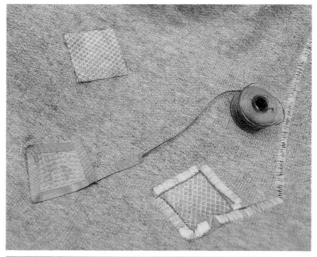

Figure 2-39 Wrap conductive patches with tape and secure thread.

sew around the edges, and then hold the tail of conductive thread against the trace you drew in step 3 as seen in Figure 2-41. (You may want to use regular transparent tape to hold the conductive thread in place, but you will need to remove it as you sew.)

Remember that the top thread and bobbin thread in your machine should not be conductive. Using an appliqué stitch (or a zigzag stitch with colorful embroidery thread), follow the template, and stitch around the bottom-right conductive patch as shown in Figure 2-40, and then couch the conductive thread against your hoodie and cover the thread with a satin stitch. Always be careful not to sew the front and back of the hoodie together. You may not be able to

continue the appliqué stitch all the way on your sleeve, so once you've appliquéd up the length of the hoodie body, you will have to hand sew your circuit with a running stitch to the conductive tape trace on the fabric switch (see Figure 2-42). This means that you will need a long tail of conductive thread for sewing. Make sure that you measure accordingly before sewing.

Once you are satisfied with the length of your appliqué stitch, release the pressure foot and remove the hoodie from the sewing machine. Cut the top thread or appliqué thread, but keep your conductive thread long and loose so that you can hand sew to the first individual fabric switch. You'll want to use a running stitch with your conductive thread and follow the circuit trace you mapped out to the conductive tape lead on your fabric switch. Once you've sewn to the tape lead, loop the conductive thread around the fabric tape three to four times (shown in Figure 2-42.) On the wrong side of the fabric, tie off your conductive thread and cut it. Repeat for every fabric switch.

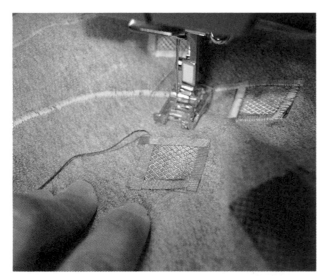

Figure 2-40 Appliqué around patch.

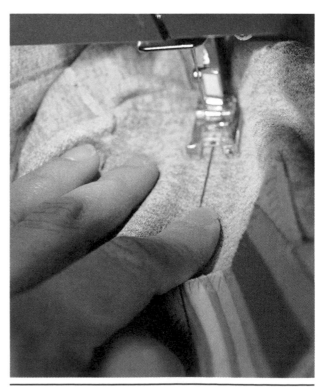

Figure 2-41 Appliqué over conductive thread.

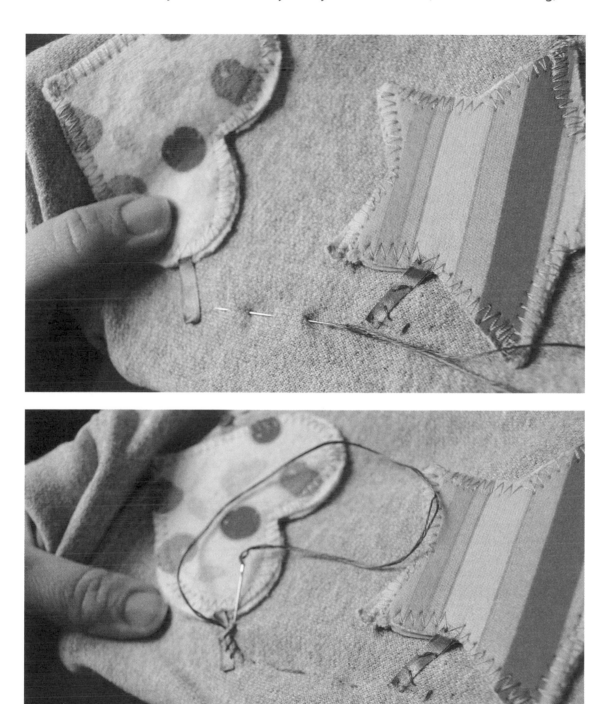

Figure 2-42 Running stitch to conductive tape leads.

Option 2: Hand Sewing

Glue conductive patches as in Figure 2-43. For this option, you will map circuitry in the same way, but since you can insulate your circuits with fabric paint, you may want to incorporate your buttons to look like musical notes (or some other illustration), as we did in Figure 2-44. In the next section we will show you how to create soft leads for hooking up your circuitry to the Makey Makey. To prepare for that, you need to sew one end of a metal snap with conductive thread onto your conductive patches, as shown in Figure 2-45. Even if you decide to use alligator clips to hook up your Makey Makey, a metal snap will give you a great place to clip your alligator clips. Plus it will make it easy to remove the electronics and allow your musical circuit hoodie to be washable!

Using conductive thread, sew the male side of the metal snap to a conductive patch, making sure to loop two times around each hole in the snap. This will ensure a connection and keep the snap in place. Do not cut the thread; instead, use a running stitch to sew along your circuit trace from one conductive patch to one fabric switch tape lead (see Figure 2-46). Once you stitch to your fabric switch, using the same thread, sew from the outside of the fabric tape into the middle of the fabric tape, as shown in

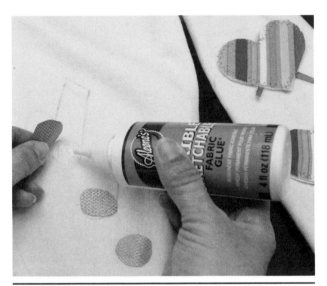

Figure 2-43 Glue the fabric.

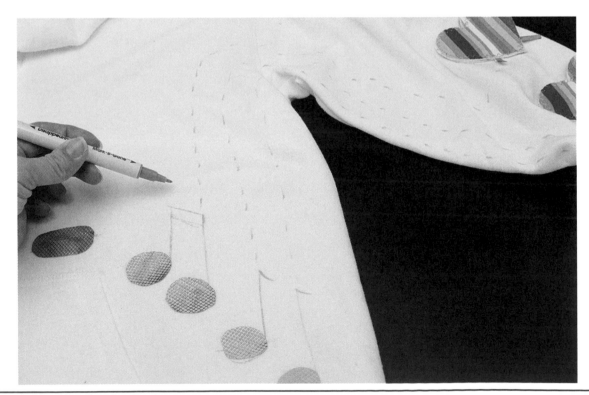

Figure 2-44 Map the circuitry traces.

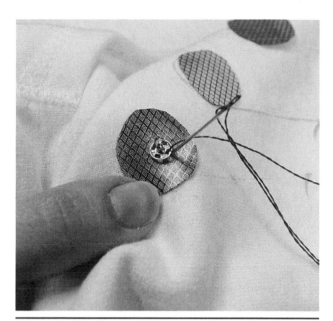

Figure 2-45 Sew metal snaps to the conductive patches.

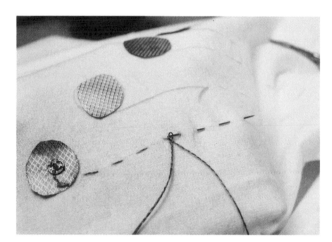

Figure 2-46 Hand stitch circuit traces.

Figure 2-47. This again will ensure a connection and help to keep the fabric switch in place. Once you've sewn three to four stitches into the fabric tape, bring your needle to the inside of the hoodie, and tie off and cut your thread close to the knot. Repeat each step for each circuit trace and fabric switch. (Refer back to the template in Figure 2-37.)

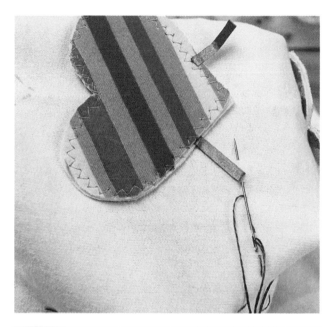

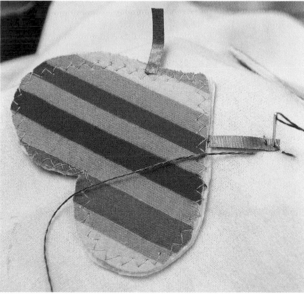

Figure 2-47 Connect traces to switch input.

Step 5: Sew Negative

Now that you've got all your key input routes sewn, it's time to sew the negative circuit trace.

Option 1: Appliqué/Couching Technique

Use fabric tape to connect conductive thread to your EARTH conductive patch as you did with your other conductive patches. Cover all four sides of your patch with fabric tape, and then hold the tail of thread against the trace you drew when mapping the circuitry. Using an appliqué stitch, sew around the edges of your EARTH conductive patch, and then, following the template, couch the conductive thread against your hoodie and cover the thread with a satin stitch. Once you are satisfied with the length of your appliqué stitch, release the pressure foot and remove the hoodie from the sewing machine. Cut the top thread or appliqué thread, but keep your conductive thread loose so that you can hand sew to the EARTH lead on each fabric switch. You'll continue hand stitching from EARTH input to EARTH input (see Figure 2-48). Following the template (Figure 2-37), sew into the fabric tape from the outside three to four times as you did when sewing the inputs. Do not tie off. Sew from the top fabric switch conductive tape lead to the next fabric switch, and sew into the fabric tape lead three to four times to connect EARTH to the fabric tape (see Figure 2-48). *Do not cut thread or tie off*; continue with the same thread to connect EARTH to fabric switches 3 and 4. When you get to your last fabric switch, loop three to four times into the fabric tape lead, bring the needle to the inside of the hoodie; then tie off and cut close to the knot. Congratulations! You've sewn all your circuits!

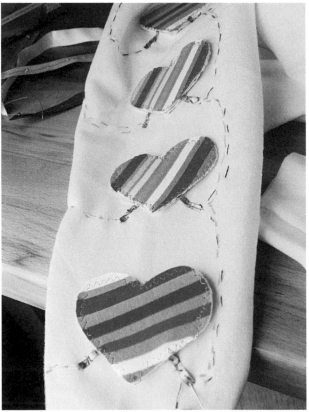

Figure 2-48   Sew from EARTH input to EARTH input.

Option 2: Hand Sewing

Sew the snap onto the EARTH conductive patch as you did with the other patches; then use a running stitch to sew to the first fabric switch tape lead. Following the template (Figure 2-37), sew into the fabric tape from the outside three to four times as you did when sewing the inputs. Do not tie off. Sew from the top fabric switch conductive tape lead to the next fabric switch, and sew into the fabric tape lead three to four times to connect EARTH to the fabric tape (see Figure 2-48). *Do not cut thread or tie off*; continue with the same thread to connect EARTH to fabric switches 3 and 4. When you get to your last fabric switch, loop three to four times into the fabric tape lead, bring needle to the inside of the hoodie, tie off, and cut close to the knot. Congratulations! You've sewn all your circuits!

Step 6: Insulate the Circuitry

Option 1: Appliqué/Couching Technique

On the inside of your hoodie, you will need to fuse very flexible interfacing to where your hand-sewn circuits are so that you do not set off stitches with your own conductive skin! Follow the directions for interfacing, which generally require using a damp cloth and an iron to fuse interfacing to fabric.

Use flexible fabric glue to insulate your hand-sewn switches, clear fingernail polish, or fabric tape on the top side of the fabric. To learn more about the fabric paint technique, read "Option 2" below.

Option 2: Hand Sewing

On the inside of your hoodie, you will need to fuse very flexible interfacing to where your hand-sewn circuits are so that you do not set off stitches with your own conductive skin! Follow the directions for interfacing, which generally require using a damp cloth and an iron to fuse interfacing to fabric. You must do this before applying fabric paint on the outside of your hoodie. If you are worried that the interfacing will make your hoodie too stiff, you can insulate your threads with flexible fabric glue or clear fingernail polish.

On the outside of your hoodie, use fabric paint to cover your circuit traces and insulate the circuitry (see Figure 2-49). Be careful to only squeeze paint gently. If you'd like, you can add designs with paint as long as you don't cover up your metal snap, accidentally insulating your connection spot!

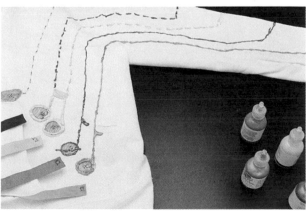

Figure 2-49 Fabric paint.

Step 7: Create Soft Leads (Or Use Alligator Clips)

Cut four colorful ribbons to 10 inches (one for each key input) and one gray or black ribbon to 14 inches (for your EARTH input.) Cut four pieces of thin hookup wire to 12-inch pieces and one piece to 14 inches. Strip one end with ½ inch of wire exposed and the other with about 1 inch of wire exposed. Thread the ½-inch exposed hookup wire to one side of the metal snap, as shown in Figure 2-50. Then use hot glue to keep the wire connected to the ribbon, as shown in Figure 2-51. Use regular thread to sew the female metal snap to the ribbon, and make sure that you have the correct side facing out so that your buttons will still snap together. (If you do it the wrong way, you can always use a seam ripper and resew.)

Figure 2-50   Hook up wire to snap.

Figure 2-51   Glue wire to ribbon.

Program and Hook to Makey Makey

Step 1: Raspberry Pi

For this project, we used a Raspberry Pi 3 with an SD card with the NOOBS software preinstalled. From the NOOBS SD card, we chose Raspbian because it has Scratch, Python, Sonic Pi, and many other software choices for educational and programming use. For more information on setting up your Raspberry Pi, visit www.raspberrypi.org/learning/hardware-guide/. You will need a USB keyboard and mouse and a monitor to plug your Raspberry Pi into to get started. Another option is to connect the Raspberry Pi to your laptop using an ethernet cable or connect wirelessly using a VNC connection. Because this is an advanced option, there are several ways to use VNC threads with your laptop on the forum boards located on raspberrypi.org.

You will want to write and save your Scratch program with your Raspberry Pi plugged into a wall outlet. Once your Raspberry Pi has booted, click the "Menu" button in the upper left, and select the "Programming" menu from this menu to open Scratch.

Create sound effects in Scratch on your Raspberry Pi. (Look at song projects in Section 4 if you want some musical ideas!) For the most part, your scripts will consist of the "When space key is pressed" block followed by the "Play sound" block. Test your sound output. If you are using an HDMI monitor with built-in speakers, it will likely come out of your monitor. We are going to use a USB rechargeable speaker with a ⅛-inch jack. So you need to tell your Raspberry Pi to switch the audio output to that source. To accomplish this, use the command line. The number 1 tells the Pi that you want to use the analog source. To change it back to HDMI, you can use the number 2.

```
amixer cset numid=3 1
```

When your program is complete, save it in Scratch. Leave the keyboard, mouse, and monitor plugged in, but switch to a portable USB power source. Navigate to your Scratch program, and test the Makey Makey and sound. Unplug the monitor and peripherals, and you are ready to go mobile.

Step 2: Hook Alligator Clips or Soft Leads

Hook alligator clips or soft leads to each arrow key input on the Makey Makey and then the other end of soft leads to each conductive patch (see Figure 2-52). Make sure to connect EARTH to your conductive EARTH patch that has a conductive trace sewn to every fabric switch. Make sure that the metal heads of the alligator clips don't touch each other. If you are using soft leads, wrap the hookup wire around each key input, and then insulate each wire with electrical tape to keep key presses from accidentally connecting and setting off other key presses (see Figure 2-53).

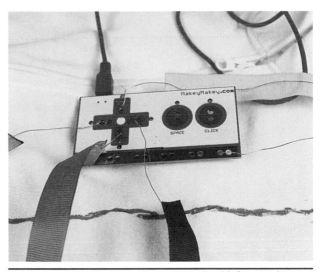

Figure 2-52 Attach soft leads to the Makey Makey.

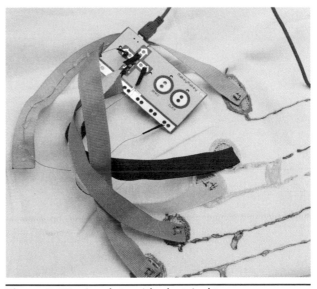

Figure 2-53 Insulate with electrical tape.

Step 3: Hide the Electronics

After programming your sounds, unplug your Raspberry Pi from the monitor, place your Makey Makey in the hoodie pocket, and find a decent-sized bag in which to keep your speaker and Raspberry Pi, as shown in Figure 2-54. Now you are ready to make some traveling beats!

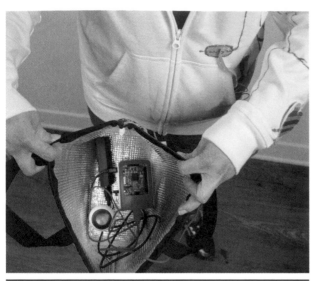

Figure 2-54   Hide the electronics.

Taking It Further

Once you learn to program one-button songs in Projects 17 and 18, you may want to come back to this project and program one of your fabric switches as the melody and then the other buttons as accompanying sound effects or beats.

Find even more ideas for wearables at http://www.kobakant.at/DIY/ or SparkFun's E-sewing tutorials here at https://learn.sparkfun.com/tutorials/lilypad-basics-e-sewing.

One-Banana Hack of Musical Hoodie: Hack a Plushie

If the Makey Makey musical hoodie seems too daunting, make a quick fabric switch, one pair of soft leads, and get ready to hack your minion's favorite plushie (see Figure 2-55).

Cost: $

Make time: 30-45 minutes

Skill level: 🍌

Supplies

Materials	Description	Source
Plushie	A minion's plushie you'd like to hack	Toy box
Conductive fabric tape, conductive fabric	Supplies from Makey Makey Inventor Booster Kit	Joy Labz
Fabric	Assorted fabric materials for your fabric switch	Joann's
Ribbon	Assorted ribbons for creating soft leads	Joann's
Thin-gauge hookup wire	50 feet of 30 AWG wrapping wire	Radio Shack (2780502)

Figure 2-55   Fabric switch plushie.

Step 1: Make a Fabric Switch from the Previous Instructions

Follow the "Make a Fabric Switch" instructions from the previous project, but you'll only need three pieces of fabric to make one switch for this one-banana hack.

Step 2: Create a Pair of Soft Leads for the Key Press and for EARTH

Make only one pair of soft leads to connect your fabric switch to your Makey Makey. Cut two colorful ribbons to 10 inches (one for key input and one for EARTH input). Cut two pieces of thin hookup wire into 12-inch pieces. Strip one end to expose ½ inch of wire and the other end to about 1 inch of exposed wire. Thread the ½-inch exposed hookup wire to one side of

the metal snap, as shown in Figure 2-50. Then use hot glue to keep the wire connected to the ribbon, as shown in Figure 2-51. Use regular thread to sew the male metal snap to the ribbon, and make sure that you have the correct side facing out so that your buttons will still snap together. (If you do it the wrong way, you can always use a seam ripper and resew.)

Step 3: Affix Fabric Switch with Glue and Hand Sew Snaps

Use stretchy fabric glue to affix your fabric switch to your plushie. Make sure that you grab the female snap if you sewed the male snap to your soft leads. Using conductive thread, connect the metal snap to the soft lead and plushie by hand sewing into the existing holes on the snap, as shown in Figure 2-56. Snap your

Figure 2-56   Sew snaps to fabric leads and plushie.

soft leads to your plushie as in Figure 2-57. Wrap one exposed end of wire to an EARTH input on the Makey Makey, as shown in Figure 2-58. Connect the second soft lead to the key press of your choice.

Figure 2-57   Snap soft leads.

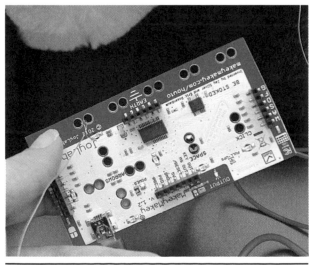

Figure 2-58   Hook to Makey Makey.

Step 4: Make a One-Button Game in Scratch

For this program, we are going to create a variable so that we can count key presses to control our plushie. Remember that a variable in Scratch is simply a way to control a value, a string of values, or even Boolean logic such as true/false statements.

Go to the "Data" palette, and click "Make a variable." In the new window that appears, type "Key press" to name your variable. We only have one sprite in this game, so you can choose either "For all sprites" or "For this sprite only."

Drag a "Set key press to 0" block to your work space, and attach it to a "When flag clicked" block from the "Events" palette. While you are in the "Events" palette, grab a "When space key pressed" block for your work area. You'll need two of these—one to control your plushie and one to stop programming if needed.

Step 5: Nest "If/Else" Statements

To create multiple sounds from one button, you will need to nest "if/else" statements that will play different sounds based on the "key press" variable. Grab two "if/else" statements and one "if" statement blocks and bring them to your work area. For the first "if/else" statement, you'll want to tell your game what you want to happen if the "key press = 0." In the "Operators" palette, you need a "□ = □" to insert between the "if" and "then" on your conditional statement. Add your desired "Play sound" from the "Sound" palette, and add a "Wait ____ secs" if needed. To increase the value of your variable, add a "Set key press to □" underneath your sound. Set your key press to turn to "1" after the first press, as shown in Figure 2-59.

Inside the first "else" section of your conditional statement, nest your second "if/else" statement to create the second sound that

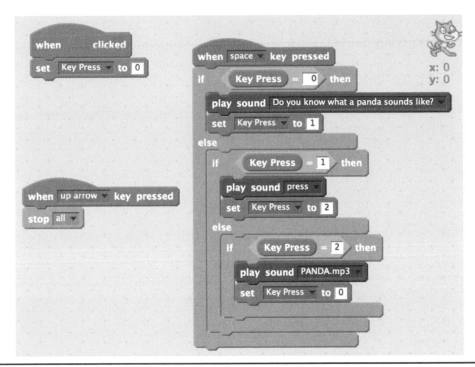

Figure 2-59 Scratch program.

will play when the key press = 1. Lastly, add your "if" statement inside the second "if/else" statement to play the third sound. You can repeat this as many times as you'd like, but make sure that your last key press will reset the "key press" variable to zero. Refer to Figure 2-59 for the full program.

Taking It Further

This is a great project to come back to after you learn the "play next" trick with variables in the Makey Go Chopsticks Project number 17. In fact, this is a fun thing to come back to after all your Makey Go tinkering because this is a one-button project!

Makey Makey Swipe Input

SOMETIMES AN EVIL GENIUS WANTS to control things with gestures. This swipe input project will get your evil genius gears turning on how to control your Scratch game with gestures and movements (see Figure 2-60). Thanks to Jay Silver for the project idea and teaching Colleen how to write this program in Scratch!

Cost: $

Make time: 30 minutes

Skill level: 🍌🍌🍌

Figure 2-60 Finished project.

Supplies

Materials	Description	Source
6B pencils	Makey Makey Inventor Booster Kit	Joylabz
Paper	Drawing paper	Craft store
Computer and access to Scratch	A free visual programming language and online community	scratch.mit.edu

Step 1: Create a Drawing

Create a drawing that will have at least four different connections, similar to the drawings in Figure 2-60. Jay's original idea was to control a game with concentric circles, so we've made a few input drawings based on this.

Step 2: Create a List (Array) in Scratch

In Scratch, you can create a one-dimensional array in the "Data" palette called a "list." In other programming languages, you can create multidimensional arrays. To find ways to use lists to create multidimensional arrays, check out the Scratch Wiki section on "Lists."

You are going to make three lists in Scratch: "Swipe Input," "Target," and "Target2." "Swipe Input" will read gestures, and in the target lists you will input the intended target players will need to swipe to control the Scratch game (see Figures 2-61 and 2-62).

Step 3: Add to Target

The first thing you need to do is to make your lists effective by adding numbers to the "Target" list and creating an empty "Swipe Input." To begin, you want to delete all information from your "Swipe Input" list and use a "Repeat" block to "Add 0 to swipe input." Now that you have created lists, these two blocks will appear in the "Data" palette. Under your "Repeat" block,

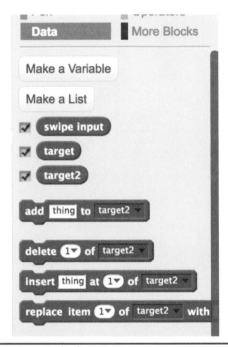

Figure 2-61 Make lists.

Figure 2-62 Replace item scripts.

drag a "Delete all of target" block, and following Figure 2-63, drag four "Add □ to target" scripts to this program. Now, in your playing field, you should see four numbers for both lists.

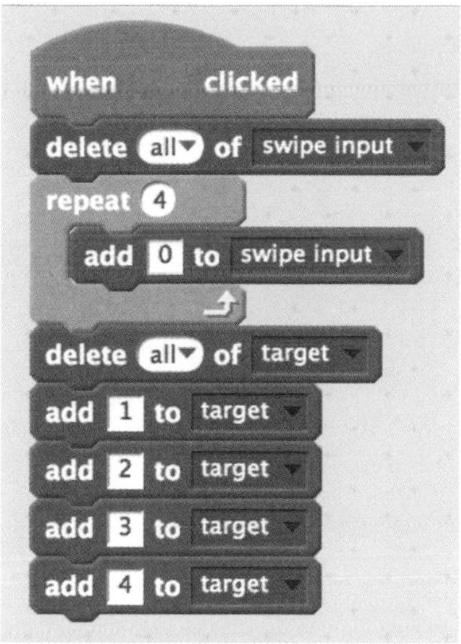

Figure 2-63 Set swipe input and target.

Step 4: Set Gestures

You will program your four arrow keys to read the gesture input. Following Figure 2-64, create a program for each key. You'll find the "Add □ to swipe input" in the data list. In order for the program to react to gestures instead of just pressing buttons, you'll also add the "Delete 1 of swipe input" block. This allows the key press to move up in the "Swipe Input" list as you touch the drawings. So if you move your hands across the keys in one direction, it will make the "Swipe Input" list match the "Target" list.

Step 5: Forever, If, Repeat

Now that your "Target" list is created and you've programmed your arrows to read the gestures, you need to tell Scratch what to do when the two lists match. Drag a "Forever" loop and "if/then" block from the "Control" palette. You want your game to be on the lookout "forever" for the lists to be matching. Drag the "Forever"

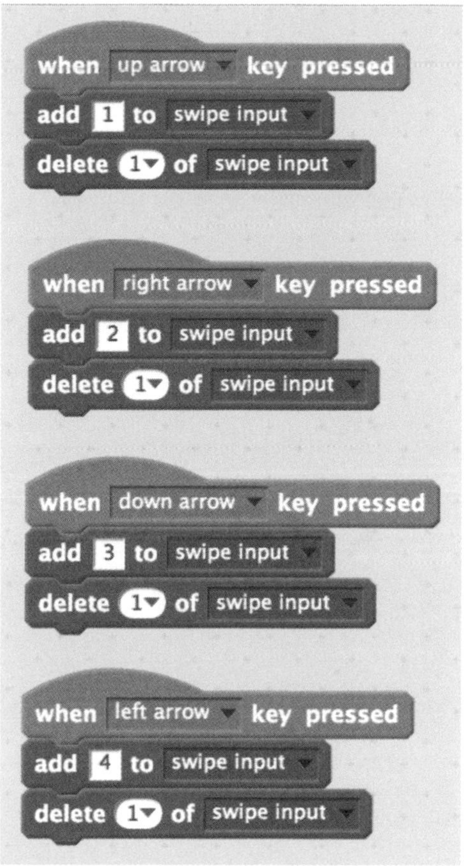

Figure 2-64 Create code for gestures.

loop to your "When flag clicked" scripts (from Figure 2-63). Place the "if/then" block inside the "Forever" loop. Now grab a "□ = □" block from the "Operators" palette, and place it in the box between "if" and "then." Add a sound effect and motion inside (or whatever action you want to occur when the lists match). Make sure that you also put four "Replace item □ of swipe input with □" blocks to reset "Swipe input" to zero. Otherwise, the action you tell the game to complete will continually glitch (see Figure 2-65).

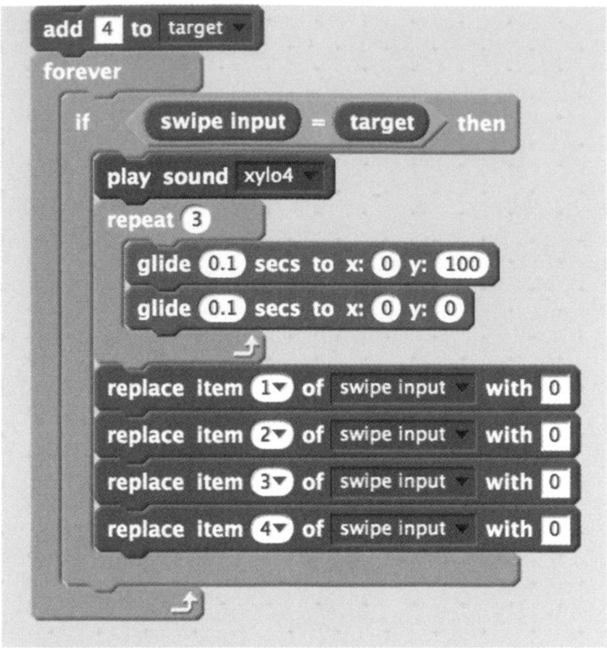

Figure 2-65 If "Swipe = Target."

Step 6: Create a Second Target

You can create a second target list that will have Scratch react with a different action by creating scripts like those in Figure 2-66. In our game, one gesture makes a sound effect play, and the opposite gesture plays the sound effect backwards. We simply took a Scratch sound, duplicated it, and reversed it in the "Sounds" tab.

Step 7: Program

Hook up an alligator clip from each drawing to an arrow key input on the Makey Makey. You may want to use a piece of cardstock or cardboard as a backing for your drawing to give it more strength.

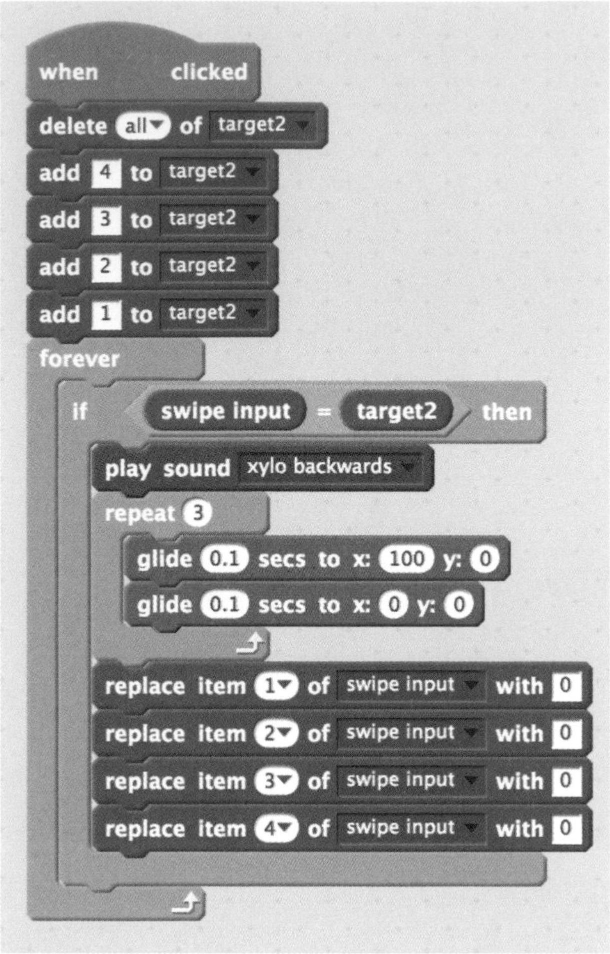

Figure 2-66 Target 2.

See the full program in Figure 2-67, and watch videos of this game in action on our book's webpage.

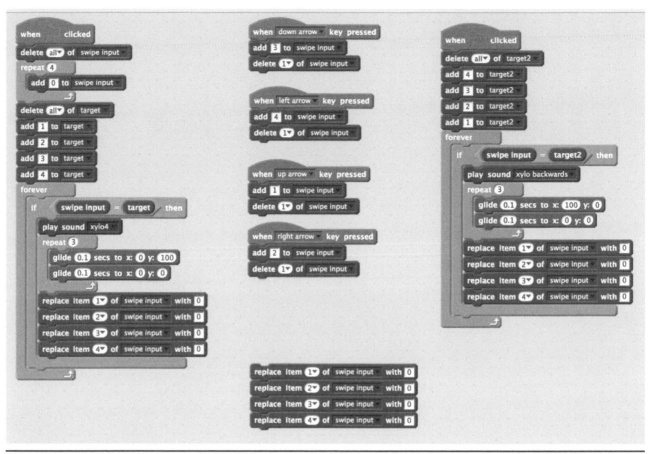

Figure 2-67 Full program.

Taking It Further

Swiping things is a great skill for an evil genius, and now you can take this swiping gesture idea and incorporate it into any program! Use it to unlock something or turn pages on your computer. Now you can scratch records in Scratch, create gesture-based wizarding spells, or any other ideas based on movements. Could you program sliding dance moves? What other conductive materials could you use to construct your swipe input?

SECTION THREE

Hacks and Pranks

Good pranks are the foundation of evil doing. Laughing at a minion's confusion when he or she goes to play a toy piano, only to find that it's been turned into a raging electric guitar, is just the beginning. Why stop with toys? Let's lock up some of your minions' favorite games and sit back and watch the befuddlement as they try to solve the lock combinations. This chapter is full of clever hacks and pranks that will keep your minions on their toes and pondering your next move.

- **Project 12:** Hacking a Kid's Toy with Makey Makey
- **Project 13:** One-Banana Hack: Hacking a Kid's Toy
- **Project 14:** Makey Makey Power Tail Prank
- **Project 15:** Makey Makey Lock Box

Hacking a Kid's Toy with Makey Makey

SOMETIMES AN EVIL GENIUS likes to take kids' toys and make them a little more … well … evil. In this project, learn how to open up the inside of a kid's toy, create inventive switches, and replace the circuit board with Makey Makey. Take those twinkling bedtime songs and reprogram them with Scratch to play more soothing music for your minions (see Figure 3-1).

Cost: $

Make time: 30 minutes

Skill level: 🍌🍌🍌🍌

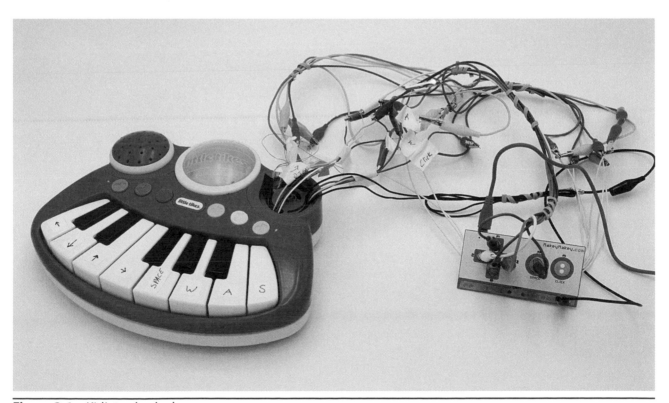

Figure 3-1   Kid's toy hacked.

Supplies

Materials	Description	Source
Toy	Electronic toy	Toy box
Conductive tape	Copper or conductive fabric tape with conductive adhesive	SparkFun or Amazon
Tools	Assorted screwdriver set, safety glasses	Toolbox
Computer and access to Scratch	A free programming language	scratch.mit.edu

Hack Keyboard Toy

Step 1: Take Out the Battery

As with any take-apart, remove the battery first to prevent getting shocked! Always use protective gear when tinkering with electronics.

Step 2: Open Up the Toy

Kids' toys often have long and cleverly hidden screws. Make sure to get a long and skinny screwdriver set designed for electronics to remove all the screws. When opening the toy, look for tabs and other things that might hold the toy together. Opening the toy carefully will allow you to evaluate how hackable it will be without breaking it.

Step 3: Investigate Easy Places to Create Switches

This keyboard toy we've opened up has buttons along the top that will make for great Makey Makey switches, and the existing keyboard keys will be fun to hack as well. For the existing top buttons on the toy, we can reprogram them with Scratch and create sound effects or use these existing buttons to change the instrument of our piano keys. While you look at your own toy, consider, "How can I use these existing buttons and switches to create simple switches with the Makey Makey?" Another thing to consider is how to create those switches without cutting

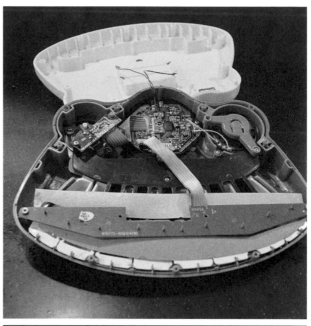

Figure 3-2   Investigating.

existing wires or rendering the toy unusable after you finish. For both of the hackable spots on our toy, there are long-running printed circuit boards (PCBs) that will make a great EARTH (or grounding) connection. On opening this toy, as shown in Figure 3-2, we can see the PCB that controls the keys, while the PCB that controls the top buttons is covered up by a large piece of plastic that holds everything in place.

Step 4: Hack Dome Switches

Cheap kids' toys typically use dome-style switches with rubber that allows the button to be pressed and then spring back up. The inside of the rubber has some type of conductive material

Figure 3-3 Close-up of original circuit traces.

Figure 3-4 Top button key press and EARTH on PCB.

(sometimes graphite, sometimes steel traces are embedded), and this conductive spot on the button completes the open circuit that is printed on the PCB when it is pressed. Figure 3-3 is a close-up of the circuit traces so that you can see how the original connection was completed once the button was pushed. Inside the top of the rubber there is conductive material that completes the connection across the circuit traces. Because the conductive material is inside, this "button" can bridge copper traces on the PCB. You can easily add a small, thin hookup wire to the top of the rubber button press by poking the wire through, as shown in Figure 3-4.

When you hook this wire to the Makey Makey, you will be able to control this surface area as a key press. Because we are switching the way this button works, we'll make each key press with these thin hookup wires and create our EARTH input along the entire PCB, which will cover up the original and individual circuit traces printed on the PCB (see Figure 3-4).

To re-create these switches, push a thin hookup wire for each button press through the rubber membrane so that it makes a connection across the inner conductive pad. Push the wire through to the other side, and make sure to twist the wire together on the outside of the rubber, as shown in Figure 3-4. Be careful not to tear the rubber as you twist the wire because we want to reuse as much of the original toy as possible. These existing button presses have a membrane inside the toy that allows the toy's buttons to spring back up after pressing (like the buttons on a calculator), so you want to be careful to keep them in working springing order.

Once you thread each wire through its button spot, run one strip of conductive tape across the PCB to create a ground for your circuit, as shown in Figure 3-4. Make sure to attach a hookup wire to your EARTH connection and label the opposite end of your wire clearly as EARTH.

Now you can place the rubber key press back over the PCB, as shown in Figure 3-5. The colorful buttons on the toy now can be programmed in Scratch and controlled via your Makey Makey hacked switches.

Step 5: Create Keyboard Switches with Copper Tape

To create switches on the existing piano keys of this toy, apply copper tape that runs from the spot where the connection will be made to the end of the piano key, as shown in Figure 3-6. Because each key will be a separate note, a separate tape trace is needed for each key. Also, to make these switches sturdier, we are going to solder our hookup wire to each tape trace. Cut 18-inch strips of hookup wire to run out of the piano toy through the old off/on switch. We used a little bit of solder to connect a wire to each piece of copper tape.

Figure 3-5 Full switches intact for top buttons.

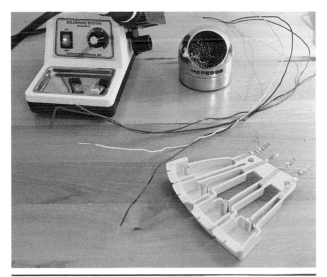

Figure 3-6 Copper tape for each key press.

Once you get your keys in place, you may need to insulate the switches to keep them from constantly making connections. Insulate as needed with small pieces of paper, as shown in Figure 3-7. It is important when using this many wires to clearly label each key press on the copper tape and at end of the wires, as shown in Figure 3-8. (An alternative method would be to connect these wires through a ribbon cable and use color coding. You will still need to clearly label each key press.) To create EARTH for these switches, run one continuous piece of copper tape along the top of the existing button presses, as shown in Figure 3-9. (Remember that these

Figure 3-8 Label wires clearly.

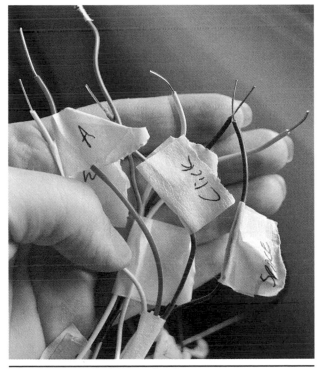

Figure 3-9 EARTH for piano keys.

rubber button presses help the keys to spring back up.) To keep the copper tape in place and not create a short accidental press from loose copper tape, we wrapped electrical tape between each button press.

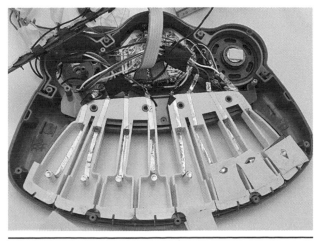

Figure 3-7 Insulate if needed.

Step 6: Put It Back Together

Because our minions still like to play with this toy, we kept the original wiring in place and only removed the off/on switch so that we could put it back later if needed. Figure 3-10 shows the toy put back together with our newly hacked switches in place. Hook alligator clips to each wire, and match labels to inputs on your Makey Makey.

When you are hacking an existing toy like this, you'll probably have to debug a few switches. This is a good thing! It will teach you more about creating your own switches and how the switches are constructed in electronic toys and gadgets. Be careful when putting screws back not to screw them too tightly because it may make your connections constantly go off. While you test each key, make sure that you clearly label each key press on the toy. You can use small pieces of masking tape to debug your switches.

Figure 3-11   Debugging tips.

(See how we labeled a key press "stuck" with tape in Figure 3-11.) When you are flipping a toy over to create switches, it can get very confusing figuring out which key is having problems, so use this low-adhesive tape to help you debug!

Step 7: Remap Makey Makey Buttons

To give you more key press inputs, you'll need to remap the mouse pins located on the backside in the right-side header. Before plugging your Makey Makey into the computer, navigate to http://makeymakey.com/remap/. (Note that you have to have version 1.2 for this remap to work.) Choose Makey Makey Classic, and then, following the graphic, attach one alligator clip to the up arrow and the down arrow and another to the left arrow and the right arrow. Now you are ready to plug your Makey Makey into the USB port. Once you plug in your USB, all the keys will light up on your Makey Makey, and your screen should say, "Makey Makey detected, now remove the clips." The instructions on the site will tell you to "Hold EARTH with one hand; press the arrows with your other hand on your Makey Makey to navigate. Press 'Click' to select." Hold EARTH, and use the right arrow until you get to the mouse header pins (shown in Figure 3-12). Holding EARTH, press "Click"

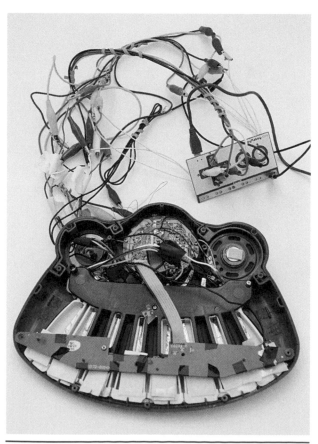

Figure 3-10   All switches in place.

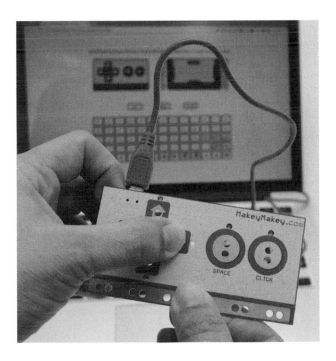

Hold Earth with one hand, press the arrows with your other hand on your Makey Makey to navigate. Press CLICK to select.

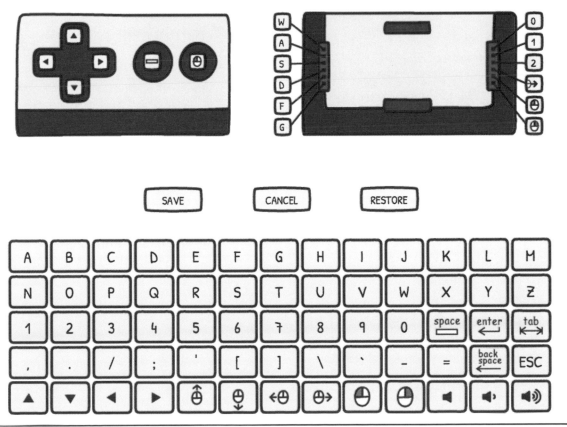

Figure 3-12   Remapping keys.

You can close this window and begin using your Makey Makey.

Figure 3-12 Remapping keys (*continued*).

with your other hand, and remap the pin from "Mouse up" to "0" by using the keys to highlight "0." Press "Click," and your Makey Makey "Remap" page should now show a "0" where it previously had "Mouse up." Use the right arrow input on the Makey Makey to highlight "Mouse Down," and press "Click." Remap this pin to the number "1." Repeat for mouse left to remap as the number "2." Once you have remapped to your evil heart's desire, press the down arrow on the Makey Makey, which will highlight "Save." Press "Click." The website will prompt you to "CONFIRM: SAVE" and is automatically on "NO." Press the left arrow input to highlight "YES," and then press "Click." You should then see the "Success" window and now be able to move to programming in Scratch. The cool thing about this remap is that your Makey Makey will hold the remap after unplugging. (It is important to note that if you remap a Makey Makey Go, it will not hold the remap after you unplug it from your device.)

Program with Scratch!

Step 1: Remix Piano Project

Look for the "Scratch Piano" by "Someone10" at scratch.mit.edu. Click "See inside" and then "remix" to save a copy that you can edit and personalize.

Step 2: Label Sprites with Key Press

To help with programming, label each key press on each sprite in the "Costumes" tab as in Figure 3-13. Use the drawing feature to draw the arrows and the text feature for labeling space, w, a, and s.

When you click on a key sprite, duplicate the lines of code for "When sprite clicked," and on the duplicated scripts, change the "Events" block to a "When space pressed" so that each keyboard key can now be clicked or played via a key press. Change each key to match Figure 3-14.

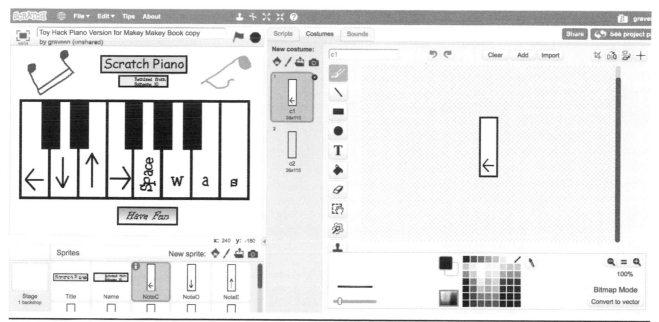

Figure 3-13   Label key presses in "Costumes" tab.

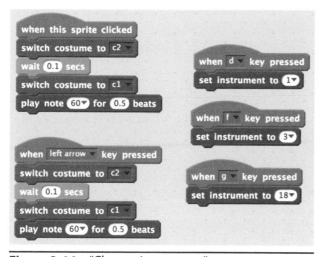

Figure 3-14   "Change instrument."

Step 3: Add "Change instrument"

To program the top three buttons on the left side of this toy so that we can change the instrument sound on the keyboard, we added the three scripts shown in Figure 3-14. The first, second, and third buttons will change the instrument sound to the chosen sound effect. All you need is a "When space clicked pressed" block located in the "Events" palette and a "Set instrument to 1" block located in the "Sounds" palette. Choose the instrument sounds you prefer, and hack Figure 3-14 accordingly. You'll also need to drag these three programs to each key sprite so that the sound effect will change for each piano key based on one button press! You can quickly copy scripts by dragging the code to the sprite in the "Sprites" menu. This will copy the code to that sprite. Do this for every piano key sprite!

Step 4: Add Sound Effects and "Stop all"

For the remaining top buttons on the right side of this toy, we added two more sound effects and the ever-important "Stop all" script. Here is where the remapping of the Makey Makey will help you to create more sound effects in Scratch. Add sound effects with the "Repeat" block, as shown in Figure 3-15, so that you can set a beat for your minions to play along with. Then make sure to program the last button with the control block "Stop all." In this way, if your minions get too loud, you can quickly use the "Stop all" block to silence the sound effects.

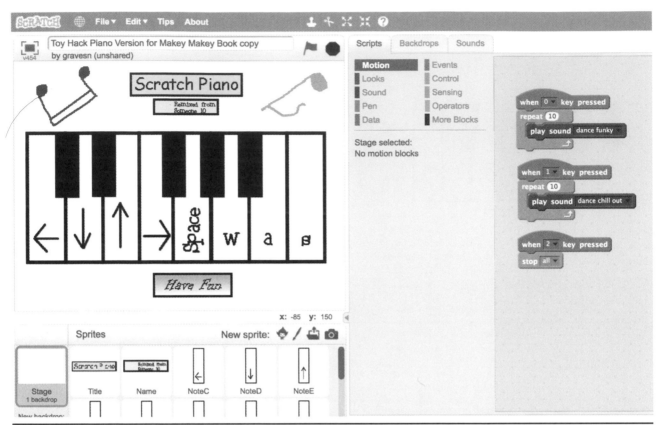

Figure 3-15   "Stop all."

Step 5: Create a Copy and Add Sound Effects

Because this is an easy remix, let's make another copy and change all the sound effects. For this second game, you will make a drum machine. Just reprogram the piano keys to be drum sound effects similar to Figure 3-16.

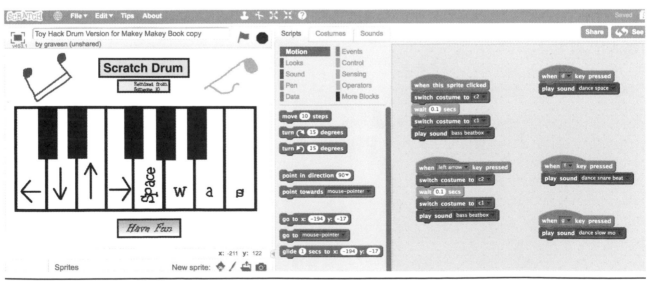

Figure 3-16   Add sound effects.

Step 6: Create Another Copy and Record Sounds

We made a third copy of this piano and recorded heavy-metal guitar riffs via the "Sounds" tab for each keyboard press. Now you can create and re-create as many sound effects as you'd like with this piano remix, which will give your newly hacked toy endless programming possibilities.

Step 7: Rotate between Scratch Games to Make Music

Let your minions rotate between Scratch games to fit the mood of the day. Now that you know how to quickly remix and hack existing Scratch games, change your music on a whim and have fun playing your hacked kid's toy (see Figure 3-17)!

Taking It Further

If you'd like to tinker further with electronics, you can look into hacking the existing circuit traces on the PCB or even tinker with circuit-bending your toy.

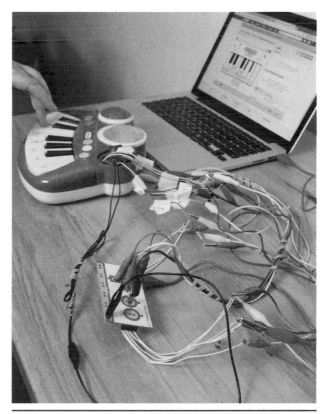

Figure 3-17   Play music.

One-Banana Hack: Hacking a Kid's Toy

An evil genius sometimes requires a quick hacking of a kid's toy. Maybe your minions would prefer a toy xylophone that doubles as an electric guitar (see Figure 3-18)?

Cost: $

Make time: 2 hours

Skill level: 🍌

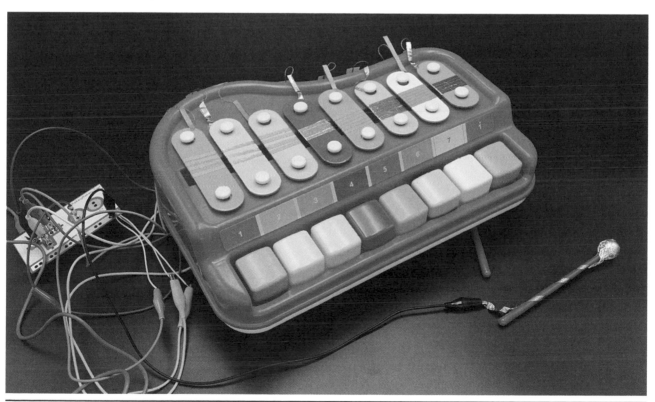

Figure 3-18 Toy xylophone hacked.

Supplies

Materials	Description	Source
Xylophone toy	Toy that uses a mallet or key and hammer to strike a metal bar to produce sounds	Toy box
Conductive tape	Copper or conductive fabric tape with conductive adhesive	SparkFun or Amazon
Tools	Small assorted screwdriver set	Toolbox
Computer and access to Scratch	A free programming language	scratch.mit.edu

Step 1: Determine Conductivity

Try using your Makey Makey to test the keys of your xylophone. Plug the USB into the computer, and hook up an alligator clip to an EARTH input and another to the SPACE input. Now, hold both alligator clips (being careful that your hands don't touch the metal on the alligator clips) to a key to see if it completes a circuit and registers a click with your Makey Makey. If it does, you can add alligator clips to each key and skip to step 5! If it doesn't register a click, chances are that your keys are coated or painted and there isn't enough bare metal for Makey Makey to register a click. It's time to open up that xylophone and quickly hack the keys.

Step 2: Open Up the Toy Xylophone

No batteries here! Open up your xylophone, and explore how you can release the xylophone keys so that you can wrap them in conductive tape. As you can see in Figure 3-19, simply squeezing the plastic that held the keys releases them. While you are here, you can also tinker with creating a switch from the button press on this xylophone.

Step 3: Wrap the Xylophone Keys with Conductive Tape

Wrap the middle of your key with conductive tape, making sure to keep the tape only in the center of the key so that it will still vibrate when

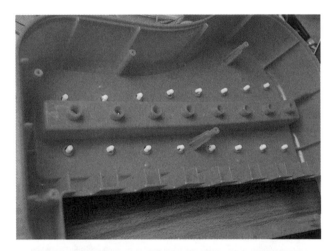
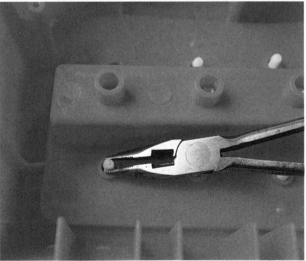

Figure 3-19   Inside of the xylophone.

hit with a mallet. If we connected an alligator clip to the key from the tape, it would decrease the vibration and tone due to the added weight. So place a small piece of conductive thread inside your conductive tape and fold the tape over, as shown in Figure 3-20.

Figure 3-20 Conductive thread to conductive tape.

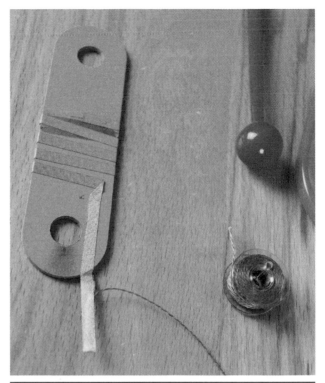

Figure 3-21 Experiment with tape wrapping and vibration.

You may want to experiment with the amount of tape you wrap around the key and see how the different amounts of tape or even types of conductive tape will affect the sound and volume of your xylophone. See a different amount of tape wrapping in Figure 3-21.

We attempted to make each metal pin (located in the holes in the center of each key) as an EARTH input so that someone could press the buttons on this xylophone and have the metal pin hit and complete a switch. Unfortunately, the metal pin hits the key at such a rapid pace that it doesn't cross the threshold for enough samples to register as a click. When creating switches with Makey Makey, both the time and area are important. If the contact is only made during

a single sample of the sampling rate, then the sampling algorithm will think it is a glitch or just noise and not an actual key press. Onc way to see this concept in action is to connect EARTH to the metal pin of this xylophone key press and an another input to one of the metal xylophone keys. Put all keys back, and try pressing the key you hacked. Does Makey Makey register an input? Now try flipping your xylophone over so that the metal pin will be in constant contact with the xylophone key. Does your Makey Makey register an input now? Cool, huh?

Because Makey Makey works better with human interaction and you want a consistent result, make your mallet into EARTH to ensure connectivity.

Step 4: Test Ring Sound

As you connect alligator clips, you'll want to test the ring sound and make sure that your keys can vibrate enough to still make a good

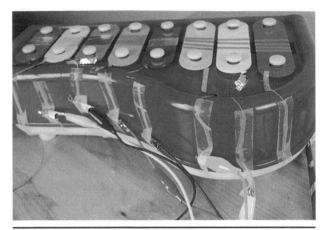

Figure 3-22   Connect alligator clips.

xylophone tone. Let your conductive thread still be somewhat loose as you tape it to the back of your xylophone, as shown in Figure 3-22. Wrap the thread around the alligator clip, and wrap nonconductive tape around the tip of the clip to hold the thread in and also insulate the connections (see Figure 3-23).

Step 5: Create EARTH with Mallet

Wrap the tip of the xylophone mallet with foil, and use copper tape to spiral around the handle so that you can alligator clip an EARTH connection to the end of the mallet (see Figure 3-24).

Step 6: Program with Scratch

Follow the directions to "Program with Scratch" in Project 12, and make as many remix pianos as you'd like. Now your one-banana hack xylophone can have as many sound effects as you can create!

Taking It Further

Now that you know the basics of transforming another instrument, start thinking of ways to transform things that weren't intended to be instruments into melodic machines. Start looking for things that would make good keys to trigger events!

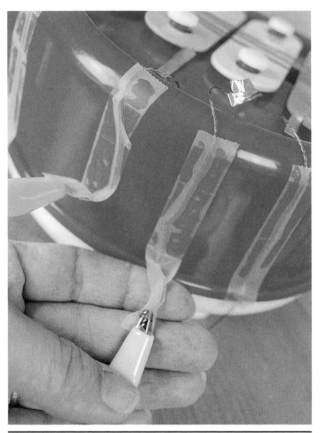

Figure 3-23   Close-up of alligator clip to conductive thread.

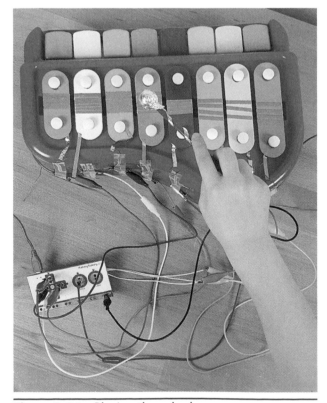

Figure 3-24   Playing the xylophone.

Makey Makey
Power Tail Prank

PowerSwitch Tail II is a great way to safely control items that are normally plugged into a power outlet with the Makey Makey. It's also a great way to make lights or a radio come on when someone triggers a switch. When the switch is released, it breaks the circuit and turns the radio off (see Figure 3-25). Although our purpose in this book is strictly evil, a PowerSwitch Tail combined with Makey Makey is a great way to create assistive technology.

Cost: $$

Make time: 30 minutes

Skill level: 🍌🍌

Figure 3-25 Power Tail prank.

Supplies

Materials	Description	Source
PowerSwitch Tail II	A device for controlling an alternating-current (AC) power switch with a microcontroller	powerswitchtail.com
Conductive tape	Copper or conductive fabric tape with conductive adhesive	MakeyMakey Booster Kit
Tools	Small slotted screwdriver	Toolbox
Computer and access to Scratch	A computer programming language	scratch.mit.edu

Step 1: Make a Plan

For this prank to work well, you need to hide the switch. We decided to make a pressure switch inside a pillow that will rest on a chair, but a simple pressure switch will also work great. Another design consideration is the devices you want to turn on when the switch is activated. A power strip plugged into the PowerSwitch Tail can switch on multiple devices. In this project, a lamp and radio will turn on when someone sits down in a nice, relaxing chair. A final planning consideration for this project is spacing. How close is the nearest outlet? How far from the outlet do you want to space the switch?

Step 2: Makey Makey to PowerSwitch Tail II

Make sure that the PowerSwitch Tail II is unplugged. The PowerSwitch Tail II has three terminals on the side, but we will only be using terminals 1 and 2. You will need a small slotted screwdriver to loosen the terminal clamps. Place one jumper wire into terminal 1 and tighten it. Run this wire to the KEY OUT output on the back of the Makey Makey. Place another jumper wire into terminal 2 and run it to the EARTH on the back of the Makey Makey (see Figures 3-26 and 3-27).

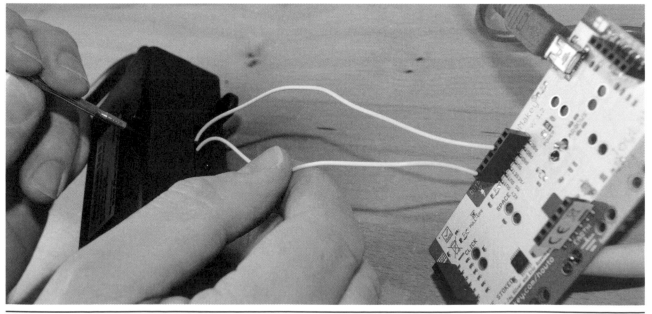

Figure 3-26 Wiring PowerSwitch Tail.

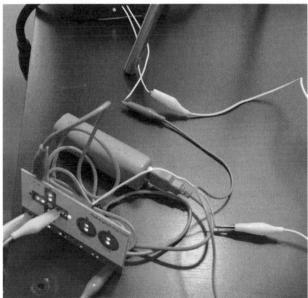

Figure 3-27   Makey Makey to PowerSwitch Tail.

Create a Pillow Switch

Step 1: Measure and Cut Fabric

To make a small envelope-style pillow case, cut two pieces of fabric to 9 by 24 inches (you can fold fabric in half and measure 9 by 12 inches). Cut four pieces of interfacing

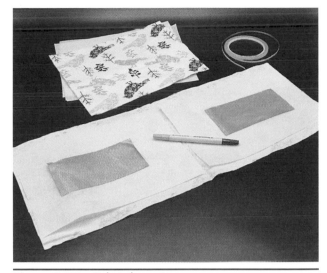

Figure 3-28   Fabric layout.

about the size of a piece of standard printer paper (8½ by 11 inches.) Cut your conductive fabric from the Makey Makey Booster Kit in half. Iron interfacing onto the fabric per the manufacturer's instructions. You will set one long piece of fabric aside (this will be the outer fabric layer and will hide your stitches in step 3), and then tape and sew the conductive pieces to the other long piece of fabric. Figure 3-28 shows the fabric layout.

Step 2: Tape Traces

Using the conductive tape from your Inventor Booster Kit, tape down one side of each piece of conductive fabric, and create two separate traces, as shown in Figure 3-29. To create neat folded edges, when you want to make a fold, simply press the fabric back with your thumbnail, and then angle the tape to make a nice "hospital corner–style" fold. Once you get to the end of the fabric, continue your tape for 4 inches, and then create a folded tab by folding 2 inches of tape back onto itself. This is where you will eventually clip your alligator clip.

Figure 3-29 Conductive tape traces.

Step 3: Wide Zigzag Stitch

Now it's time to sew your conductive fabric to the inner layer of your pillow switch. Set your sewing machine to a zigzag stitch, and widen your stitch to sew the width of the conductive fabric tape. Test your zigzag stitch width on a remnant of fabric before sewing on your final fabric. Sew around the conductive fabric and along your conductive tape trace to about 1 inch from the edge of the fused fabric, as shown in Figure 3-30. Sew both conductive fabric pieces and tape traces.

Step 4: Hem, Pin, Sew

On your sewing machine, narrow your zigzag stitch back to a straight stitch. Take your outer layer of fabric, and place it over your inner layer. Fold the ends of this pillow case in about 1 inch, and then sew a straight stitch to create a hem, as shown in Figure 3-31, making sure that your conductive fabric tape tabs are still clippable. Placing the right sides of the pillow

Figure 3-31 Pinning together sides.

case together, pin your fabric as shown in Figure 3-31. Sew the sides together with about ½ inch of seam, as shown in Figure 3-32.

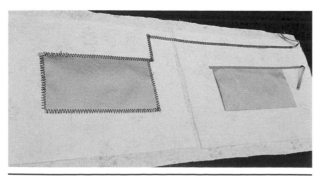

Figure 3-30 Sewing along tape.

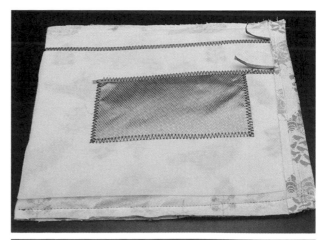

Figure 3-32 Sew sides.

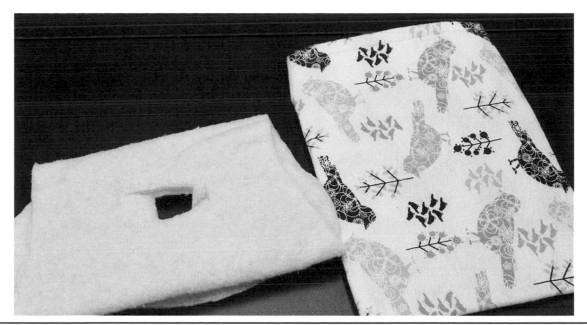

Figure 3-33 Creating an insulator.

Step 5: Insulate Your Switch

To create a barrier between your key press and EARTH, you are going to take a layer of quilting batting and fold it so that it fits inside your pillowcase. Then cut a hole in the center of the batting (see Figure 3-33) so that when a minion sits on the pillow, it will make the conductive fabrics touch and complete the circuit. Place this inside your pillow fabric, clip one alligator clip to a key press and the other to an EARTH input on your Makey Makey (see Figure 3-34). Test your switch to make sure that it works before hooking up to the Power PowerSwitch.

Figure 3-34 Complete pillow switch.

Step 6: Put It All Together

Hide your pillow switch on a chair, plug your PowerSwitch Tail into an extension cord, and plug the extension cord into the wall. If you have a USB port on your extension cord, you can plug your Makey Makey into the extension cord. Otherwise, you will have to use a USB phone charger or plug the Makey Makey into your computer. Plug in the things you would like to come on when your minions sit and test out your prank. You can see a video of this project in action on our webpage.

Taking It Further

What other ways could you cleverly disguise the switch for the PowerSwitch Tail II? Could the PowerSwitch Tail II also can be wired to turn things off instead of on? Check out the company's website for more information, and start thinking of possible uses!

Makey Makey Lock Box

AN EVIL GENIUS ISN'T WORTH his or her salt in evil if he or she can't create a complex lock to make his or her minions use their brains before playing games in Scratch. In this project, you'll take another look at lists in Scratch to create a complex lock box that you can add to the

beginning of any Scratch game to create an aspect of physical computing (see Figure 3-35).

Cost: $

Make time: 30 minutes

Skill level: 🐟🐟🐟🐟

Figure 3-35 Finished lock box.

Supplies

Materials	Description	Source
Brads or tacks	Brass brads or metal tacks	Office supply store
Alligator clips	Three extra test leads	Joylabz, Amazon
Computer and access to Scratch	A computer programming language	scratch.mit.edu

Create a Lock Program in Scratch

Step 1: Create Lists

The creation of this project relies on making lists (or simple arrays) in Scratch like you did in Project 11, although this time you are going to add a little problem solving to the project. Your minions will have to decipher the numbers on the lock and how to crack it. This is physical computing at its finest! Begin by creating two lists in the "Data" palette: "Lock box" and "Open sesame." Depending on how you want to incorporate this lock, you can create these scripts in a sprite or on a backdrop of your game. If you eventually want to add this to a different game, you'll want to create these scripts on a backdrop located in the "Stage" area of Scratch.

Every time the game starts (or when the green flag is clicked), you'll want to clear both lists and set the "Lock" list. (If you don't clear the "Lock" list, it will add lines to your array rather than restart it, and you don't want this!) To do this, drag two "Delete 1 of □" blocks to your "When flag clicked" block and change the number value to "All" and the second value to each corresponding list, as shown in Figure 3-36. To set your combination lock, add three "Insert □ at 1 of lock box" blocks, and set your lock to match Figure 3-36 or set your own lock combination. Make sure that you change the script with the dropdown menu to read "Open

Figure 3-36   Create lists in Scratch.

sesame" to create the answer key to your lock box. To create an empty list for your lock box, use a "Repeat" block from the "Control" palette, and insert three "Add □ to lock box" blocks inside the "Repeat" block, changing the value from 10 in the repeat block to the number of lines you want in your lock array.

Step 2: Program Keys

Now that you have your lists set to create when you click the green flag, you need to program your keys so that they will work with Makey Makey. For this lock box, you want three rows with three brads each. These nine brads will create numbers in Scratch when a minion touches them. The fourth row will have only one brad, and it will function as the EARTH input.

To make this lock more of a puzzle, you'll make your minions decipher the numbers. The clearest way to do that is to have the first row of brads be the input for the first line in your list. The second row of brads will be the input for the second row and the third row for the third item in the list.

To program this in Scratch, you need a "When key pressed" block from the "Events" palette, an "if/else" statement from the "Control" palette, a "Key ☐ pressed?" block from the "Sensing" palette, and two "Replace item ☐ of lock box with ☐" blocks from the "Data" palette. You'll want the up arrow to replace the first line of the lock box with the number you assign. To make this lock even more puzzling for your

minions, you can also make your minions add the numbers on the lock to break the lock box. To do this, you need to tell Scratch to make the number 3 appear as the first item in the list only if the up arrow is pressed along with the down arrow. Follow Figure 3-37 to create your first row of numbers for your lock box. To make the numbers appear for the second item in the list, follow Figure 3-38. Now your minions can only unlock the lock box if they can figure out how to add the unlabeled numbers and make line 2 display the number 9. The scripts for the third row are programmed as shown in Figure 3-39. Once you feel comfortable creating this program, see how far you can take the number lock combos!

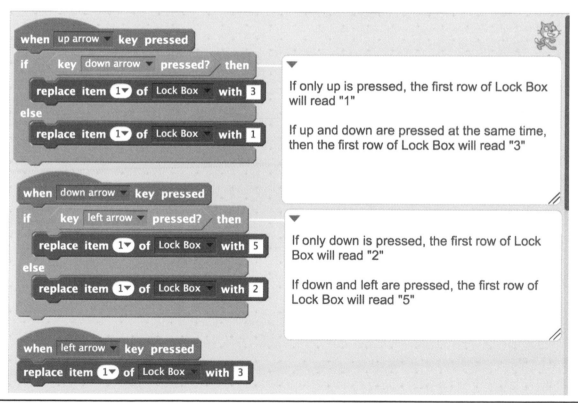

Figure 3-37 First row of lock box keys.

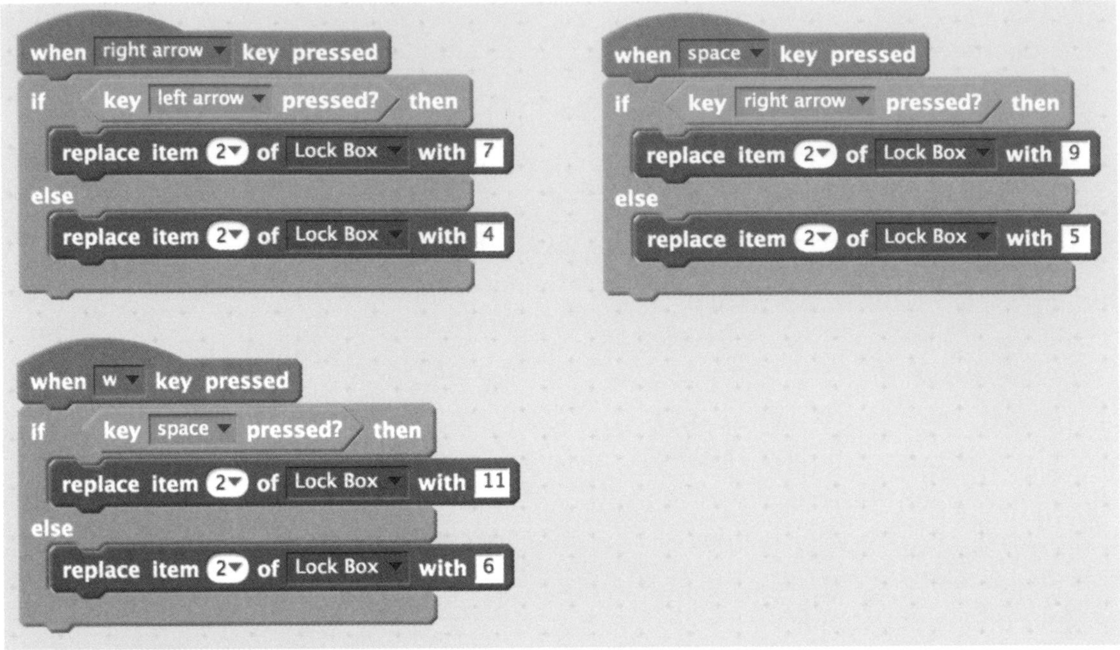

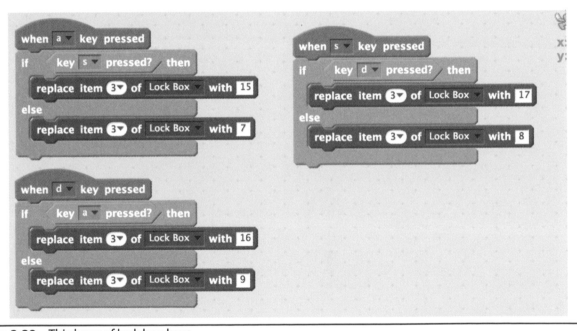

Figure 3-38 Second row of lock box keys.

Figure 3-39 Third row of lock box keys.

Step 3: If This, Do That

Now that your keys work, you'll need to tell Scratch how to let your minions know when they've unlocked the lock box. For this we need a "Forever" loop and an "if" statement from the "Control" palette. Nest the "if" block inside the "Forever" loop, and then grab a "☐ = ☐" green block from the "Operators" palette. Insert the operator in between "if" and "then." Then grab your list's names from the "Data" palette. Make sure that your code looks similar to that in Figure 3-40. Inside the "if" block you can insert

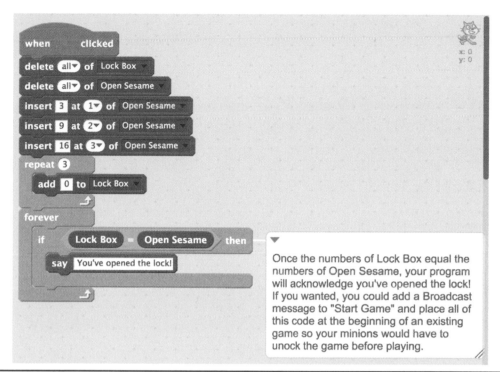

Figure 3-40   If lists match.

whatever behavior you want to happen when your minions unlock the code. We put a simple "Looks" block that reads, "You've opened the lock." You could also insert a "Broadcast" message so that your minions could play a game after unlocking the box. Or you could hide a secret message, interesting sound bite, whatever your heart desires!

Create a Lock Box

Step 1: Mark Brad Placement

Once you find the perfect box for the lock box, you'll want to measure and find the center of the box and then mark your other brads about 1.5 to 2 inches apart depending on the width of your box. Use a box cutter to cut a small slit for each brad placement, and then push the brads through. You should have three rows of three and then one brad for EARTH (see Figure 3-41).

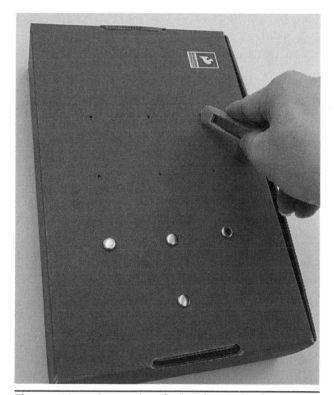

Figure 3-41   Create slots for brads.

Table 3-1 Brads to Makey Makey from the Inside of
the Lock Box

3	2	1
Left	Down	Up
arrow	arrow	arrow
6	5	4
W key	Space	Right
		arrow
9	8	7
D key	S key	A key

Step 2: Label Keys and Hook Up Makey Makey

On the inside of your box, label each brad to match your code, as shown in Figure 3-42 and Table 3-1. You could fold the brads out, but they actually make a great place to hook your alligator clips to. Make sure that you match the correct inputs to the correct brads and hook your clips so that they hold the brad as in Figure 3-43 and go to the correct key press as labeled in the tables.

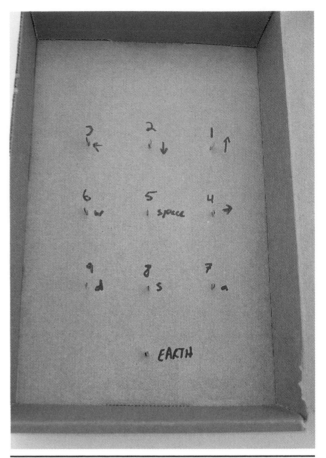

Figure 3-42 Label each brad inside the lock box.

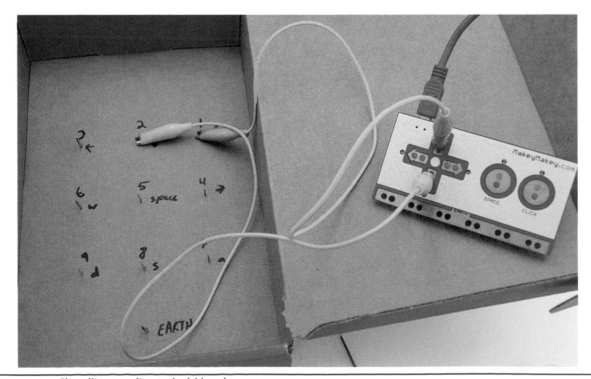

Figure 3-43 Clip alligator clips to hold brads.

Step 3: Get Ready to Play!

Once you hook all the alligator clips to your box as shown in Figure 3-44, secure any spots with electrical tape that might create connection interference, place your Makey Makey inside the box, and close it. Snake the USB cable out of the box as we did in Figure 3-45, or cut a small hole for the USB cable. Plug the USB cable into your computer and get those minions thinking! (Table 3-2 shows the key press references for each brad from the front of the box.)

Table 3-2 Brads and Makey Makey Inputs from the Front of the Lock Box

1	2	3
Up arrow	Down arrow	Left arrow
4	**5**	**6**
Right arrow	Space	W key
7	**8**	**9**
A key	S key	D key

Figure 3-45 Fully enclosed and ready to play.

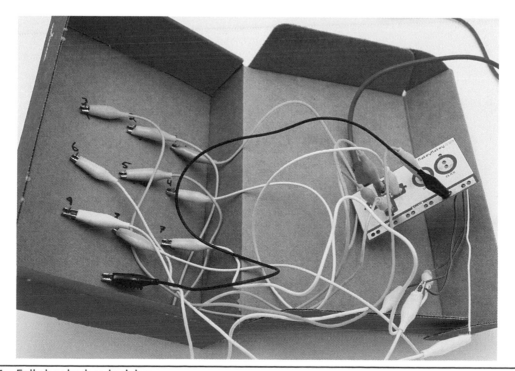

Figure 3-44 Fully hooked up lock box.

Taking It Further

What if you had to exercise before you could eat a cookie? Change your locking combo to an exercise combo, and hook up your cookie jar alarm. Make your minions complete the exercises, and then reward them with cookies.

SECTION FOUR
Makey Makey Go

Every evil genius has a utility belt full of tools to get the job done efficiently anytime and at anyplace. Even though the Makey Go is little, it is big on fun and possibilities. These projects will help you get to know the ins and outs of Makey Go and teach you some clever hacks that will deliver a large dose of evil in a small package!

- **Project 16:** Makey Go No Donut Prank
- **Project 17:** Makey Go "Chopsticks"
- **Project 18:** Makey Go "Heart and Soul" Plant Kalimba
- **Project 19:** Makey Go Cat Clicking Game
- **Project 20:** Makey Go Lemon Squeezy

About Makey Makey Go

The Go differs from the original Makey Makey because it has only one activation point, and it uses capacitive sensing. It has three buttons on it. The first is a "gear," which allows you to switch between inputs. The default inputs are "click" and "spacebar." When Go is in "spacebar" or "key press" mode, the tail end is red and flashes green when activated. When Go is in "click" mode, the tail end of Go is blue and flashes green when activated. You can remap "spacebar" to another key, but "click" will always be "click." The middle button on the Go is the "Play" button. Every time you hook up a new object to Go, you'll want to recalibrate with the "Play" button. When you tap "Play," the Makey Makey Go reads the capacitance of the object hooked up to the end of your alligator clip. The third button is the "Plus" button, which is waiting for a touch so that the Go will activate. This is where you attach your objects via an alligator clip.

To use the Go, just connect an alligator clip into the "Plus" button slot and connect the other end to something conductive. When a bigger capacitor comes along and touches the Go or the object attached to it, the Makey Makey Go realizes, "Oh, the capacitance just went way up!" and tells the computer that the "spacebar" or "click" has been pushed. For example, a glass of water is a capacitor of a certain size, but when you touch the water, you join your body's capacitance with that of the water bottle (and you are one rather large bucket of electrons!). This interaction will tell the Makey Makey to "go."

Because the Go relies on capacitive sensing and not the completion of a circuit, you can make inventions in a completely different manner. Capacitance can even go through objects, so you can put your alligator clip into a water bottle and squeeze the bottle to activate the Makey Makey Go. These last five projects will play around with this capacitive sensing and one-button hacks for Scratch to take your Go projects even further than you imagined!

NOTE: If you hold the "gear" button down for 5 seconds and your Go tail light turns white, it means that it is in "Remap" mode. Simply unplug it from the computer, and plug it back in to reset it.

Makey Go No Donut Prank

Minions love donuts, but do they really deserve one? Probably not without a little bit of a hassle, and this prank will make anyone who tries to take a donut second-guess their decision (see Figure 4-1).

Cost: $

Make time: 30 minutes

Skill level: 🍌

Figure 4-1 Completed donut trap.

Supplies

Materials	Description	Source
Conductive serving tray	Pizza pan, metal serving tray, or aluminum foil sheet	Kitchen
Conductive tape	Makey Makey Booster Kit	Joylabz
Donuts	Chocolate frosting with sprinkles	Donut store
Computer and access to Scratch	A free visual programming language and online community	scratch.mit.edu

Step 1: Make a Hassle

To start, you will need to open a new game in Scratch. To hassle your minions about taking a donut, you will make six recordings of donut warnings and alerts. Then create a program to randomly play your recorded sounds when a donut is removed from the serving tray. Begin by uploading a donut sprite from the Scratch library. Change the icing to your favorite color in the "Costumes" tab. Add as many donuts as you'd like to your stage. Then, while your first donut sprite is selected, click over to the "Sounds" tab and make a plethora of donut hassle recordings (see Figure 4-2). The more

sounds you create, the more random your prank will be!

Step 2: Make It Totally Random

Now it's time to create a short script that will randomly choose a recording when a donut is grabbed. Create a "donut" variable in the "Data" palette. Set your "donut" to zero when the game starts. Then start a new set of scripts with a "Key press" block and a "Play sound until done" block. In the "Operators" palette, you need a "Pick random 1 to 10" block that you will insert into your "Play sound" block. To have "donut" recordings play, grab the "donut" value from the

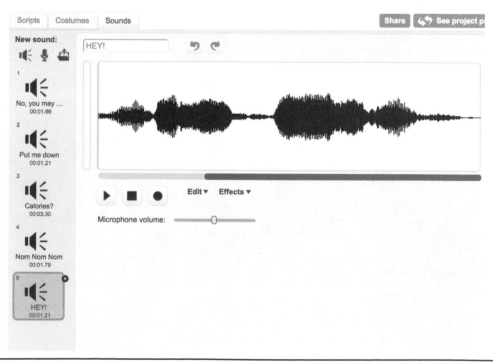

Figure 4-2 Donut hassle recordings.

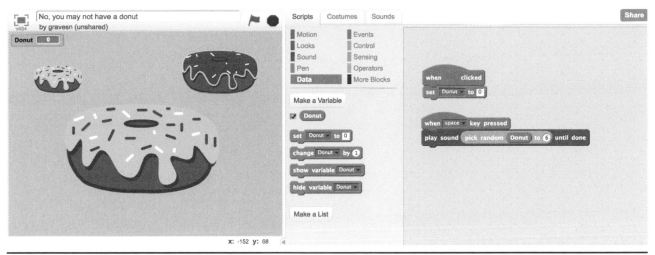

Figure 4-3   Scratch program.

"Data" palette and insert it into the first value slot. Your second value should be the number of recordings you have. In Figure 4-3, we only had six recordings, so our second value is 6.

Step 3: Set the Trap

Grab a large metal pan or serving dish that is conductive. Arrange some fresh donuts, preferably chocolate with sprinkles, on the dish. Then plug in Makey Go and clip the alligator clip from the plus sensor to the conductive serving dish. Place a tea towel next to the dish to disguise the test lead if the connection is too obvious. Be sure that you are not touching the dish, and press the "Play" button to calibrate the Makey Go.

Step 4: Test and Watch

Start the Scratch program by pressing the green arrow. Now just try to take one of those delicious donuts away, and prepare to be hassled. If removing a donut does not trigger the Makey Go, try recalibrating or switching to "sensitive" mode by pressing and holding the "Play" button for 2 seconds.

Taking It Further

People love to take things, especially if they are frosted or shiny! What other things could you hassle people about taking, and how could you cleverly hide your trap? Is there a way to make this mobile?

Makey Go "Chopsticks"

THE EVIL GENIUS LOVES SUSHI and music, so why not combine the two? In this project you will learn how to use Makey Go to play a note of a song when you dip your sushi in some soy sauce (see Figure 4-4). This is a great project to try when any kind of dipping is involved. Plus it is a splash at parties.

Cost: $

Make time: 30 minutes

Skill level: 🍌🍌

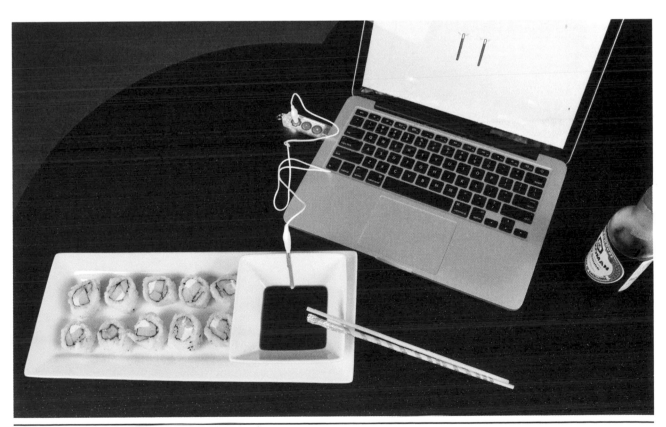

Figure 4-4 Completed Makey Go chopsticks.

Supplies

Materials	Description	Source
Flatware	Sushi platter and dipping bowl	Kitchen
Conductive fabric tape	Makey Makey Booster Kit	Joylabz
Sushi and chopsticks	California roll	Sushi restaurant
Computer and access to Scratch	A free programming language	scratch.mit.edu

Create Scratch Program

Step 1: Create and Set Variables

Pick two sprites that will function as each piano hand of your song. Click on your first sprite (which you will program as the left hand), and in the "Data" palette, create the variables "left hand" and "right hand." Create a program to "Set left hand to 1" when the green flag is clicked. Click on your second sprite (which you will program as the right hand), and create a program to "Set right hand to 1" when the green flag is clicked.

Step 2: Add Sounds in "Sounds" Tab

It's time for the piano fun! What you are going to do is upload all the piano notes needed to play "Chopsticks" in order and edit each piano key so that it sounds like playing the song "Chopsticks" in real life. No, you don't have to record all the piano keys yourself; instead, Scratch has piano keys preloaded into the "Sounds" menu. Click back to your first sprite (the "left hand" sprite), and get ready to program.

In the "Sounds" tab, click "Upload," and then choose the "Musical notes" menu. The first note of the left hand is "F piano." Click on it, and while you are here, edit the sound so that it plays like a staccato note, as shown in Figure 4-5. Just

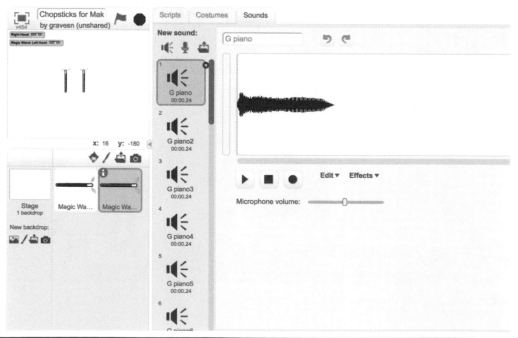

Figure 4-5   Staccato F note edit.

click and drag your mouse to highlight the audio you want to delete, and click "Delete." Test your sound to make sure that you like the length.

Now, here's the great thing: instead of repeating that step, you can just right-click on the sound and select "Duplicate." Scratch will even number the next note as "F piano2." Repeat this so that you have six "F piano" notes, and then you are ready for the next sound (see Figure 4-6). Number the sounds in your "Scratch" tab to help you remember. Each short note repeats six times until you get to the C note for both hands.

Click "Upload" > "Musical notes" > "E piano," and repeat the "E piano" key six times.

Follow the chart below to create the whole song for the left hand. Once you have all left-hand notes uploaded, click on your second sprite, and create the notes for the right hand in your second sprite. (This will allow both notes to play on one key press.) Remember to leave your C piano notes long to mimic the real piano play of "Chopsticks."

Left Hand	Right Hand
F (*6)	G (*6)
E (*6)	G (*6)
D (*6)	B (*6)
C (*4)	C2 (*4)
D (*1)	B (*1)
E (*1)	A (*1)

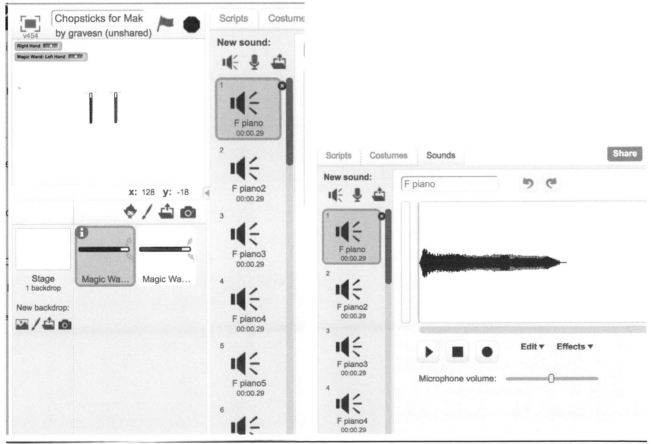

Figure 4-6   F note duplicates.

Step 3: Play Sound and Change Note (Left Hand)

You have created your notes in Scratch, but you still have to tell the computer when to play them. This is a quick group of scripts to create because we already created counter variables.

When the spacebar is clicked, you want the Scratch program to play the "left hand" notes. Simply take the "left hand" script from the "Data" palette, and insert it into the "Play sound" block as in Figure 4-7. Then add the "Change left hand by 1" block so that Scratch will count each note and play each note. It is imperative that you have the sounds in order in the "Sounds" tab or the song will not progress in the right intervals. We used the "Next costume" block so that we could change the costumes of our sprites and see each piano note play. Add a costume in the "Costumes" tab if you want to use this block.

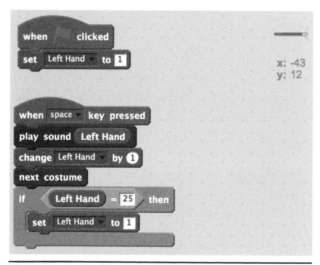

Figure 4-7   "Left hand" program.

Step 4: Play Sound and Change Note (Right Hand)

You can duplicate the script above in the "right hand" sprite by dragging the full program to the sprite. Just make sure that you change all references in scripts from "left hand" to "right hand" (see Figure 4-8).

Figure 4-8   "Right hand" program.

Step 5: If Counter Reaches Last Note, Reset to Beginning

You don't technically have to do this last step, but it will help if you program another song and need to debug. Plus, this is the same way you would program a song in Arduino based on a "sensor" or "key press." (See the Arduino stuffie project in our *Big Book of Makerspace Projects for the Arduino* version of this method with the song "Iron Man.") To get your song to return to the beginning and start recounting, use the "If" block, and nestle the operator "□ = □" between "if" and "then" so that it reads, "If left hand = 25, then set left hand to 1" (see Figure 4-7). Repeat this for the "right hand" sprite and now you are ready to play "Chopsticks" with chopsticks!

Alternatively, you can also use the scripts "If left hand > 24, then set left hand to 1." If you were writing a song in Arduino, you'd use something like the following code. This would tell the IDE to reset to zero only if the note count is higher than 24. Because Scratch doesn't have a "Greater than or equal to" block, you can just use the "Greater than" symbol to emulate this code.

```
{
if(lastnote >= 24)
{
 lastnote = 0;
 delay(100);
}
```

Conductive Chopstick and Dipping Bowl

Step 1: Create Conductive Chopstick and Dipping Bowl

With the programming work out of the way, it is time to make a conductive chopstick. For this, you will simply need to break apart some chopsticks and wrap conductive fabric tape or copper tape from the bottom to the top of one of them. If you are not eating sushi and are using carrots, sliced apples, or celery to scoop your dip, you can skip this step.

Place a piece of conductive fabric tape on the edge of the inside bottom of a small dipping bowl. Run the tape up the side of the box, and cut it about 2 inches above the top of the bowl. Fold the tape over to create a 1-inch tab. Fill the bowl with soy sauce or dip. (If you do not have conductive tape, you can always use aluminum foil and double-sided sticky tape to adhere it to your bowl.)

Step 2: Dip It Real Good!

Insert Makey Go into a USB port on your computer and clip the alligator clip from the plus sensor to the tab on the bowl. Press the "Play" button to calibrate the Makey Go. Grab the chopsticks, and dip a piece of sushi in the sauce bowl. Each time you dip, your sushi song will advance (see Figure 4-9).

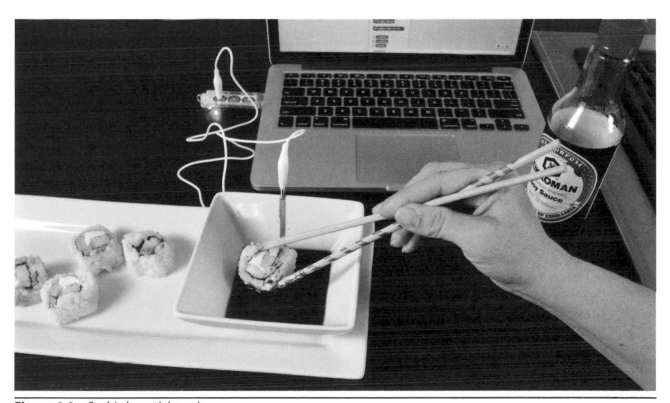

Figure 4-9 Sushi chopsticks go!

Taking It Further

As the dip or sauce runs out, will the Makey Go stay calibrated? What other songs can you program and match to their dish? How can you apply this to a buffet or ice cream sundae party?

Makey Go "Heart and Soul" Plant Kalimba (Thumb Piano)

THERE IS NOTHING MORE ANNOYING than hearing your favorite song played poorly by an adoring minion. You could take the easy way out and tell the minion that he or she is terrible, but is that truly evil? What if you could make the minion think that he or she is an amazing musician and actually create an instrument that only uses one button to play an entire song. There is no way the minion could hit a bad note, and when he or she shows off to others, it looks

like you have attracted a high-quality entourage! For this project you are going to create a simple thumb piano with house plants, a discarded box, and a programming trick in Scratch (see Figure 4-10).

Cost: $

Make time: 30 minutes

Skill level: 🍌 🍌 🍌

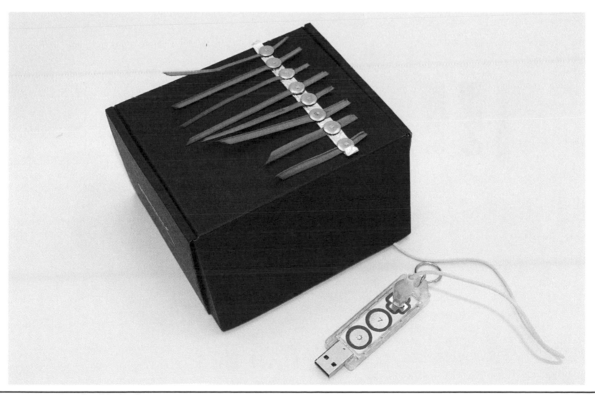

Figure 4-10   Finished kalimba.

Supplies

Materials	Description	Source
Cardboard box	A thumb piano–sized box	Recycling
Conductive stuff	Thumbtacks, aluminum foil	Junk drawer
Plant leaves	Long, skinny leaves	Nature/backyard
Computer and access to Scratch	A free visual programming language and online community	scratch.mit.edu

Program Tones

You learned one way to program a song with variables in the last project; now you are going to go further with this concept by counting notes and using the power of combining "if" statements with "or" operators to play any song you want—all with just one key press.

Step 1: Play a Song

We'll show you how to create the classic "Heart and Soul," but once you get the hang of creating music with math, feel free to create your own music with this one-button hack.

To figure out the note count for a song, look at Figure 4-11, where we've counted the notes on a keyboard for "Heart and Soul." In Scratch, you can choose a note and its duration with the "Play note 60 for 0.5 beats" block and adjusting the note sound by the dropdown arrow and adjusting the length of the beat. This is actually very similar to creating musical notes for piezos with the Arduino programming language. In the Arduino IDE, you would define the notes and note duration and then have the song play based on counting notes (also similar to the last project!).

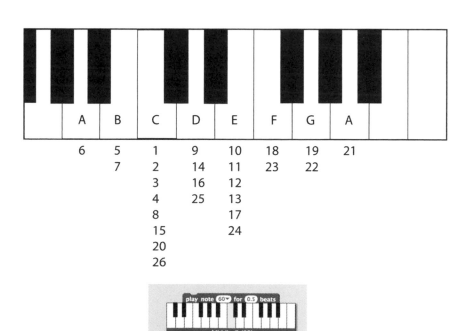

Figure 4-11   Keyboard template with "Heart and Soul" note count.

For "Heart and Soul," we drew a small keyboard similar to Figure 4-11, and then wrote down a number for each song note. You can create your own song with this same project by using the keyboard in Figure 4-11 and renumbering the keys for the song you want to play.

Step 2: Create Variable and Set Instrument

Create a variable called "next note" in the "Data" palette, and add "Set next note to 0" to a "When flag clicked" block. In the "Sounds" palette, find a "Set instrument to □" block, and pick your favorite instrument sound (see Figure 4-12).

Step 3: "If" and "Or"

Now that you have your song notes by number, it's time to double up your "or" operators and tell Scratch which note to play following your count. Every time the space key is pressed, you

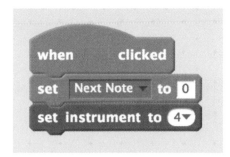

Figure 4-12   Set "next note" and instrument sound.

want the variable to change by 1. To do this, simply add a "Change next note to 1" block under the event block "When space key pressed."

To play "Heart and Soul," you need to program Scratch to play middle C when the space key is pressed only if the count is on 1, 2, 3, 4, 8, 15, 20, or 26. You can use multiple "or" operators to put all these numbers in one "if/then" statement or break them up into separate "if/then" statements, as shown in Figure 4-13. Follow the template (Figure 4-11) and Figure 4-13 to have Scratch count notes and play the

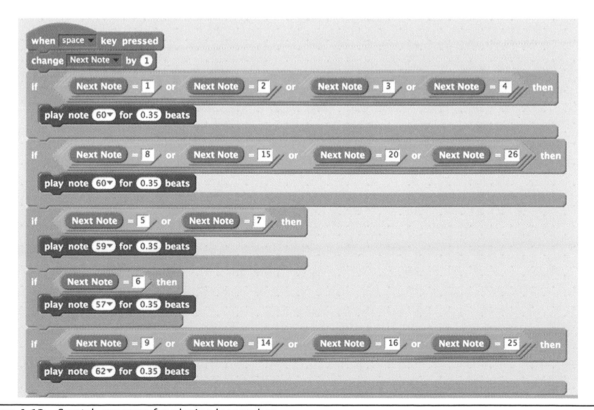

Figure 4-13   Scratch program for playing by number.

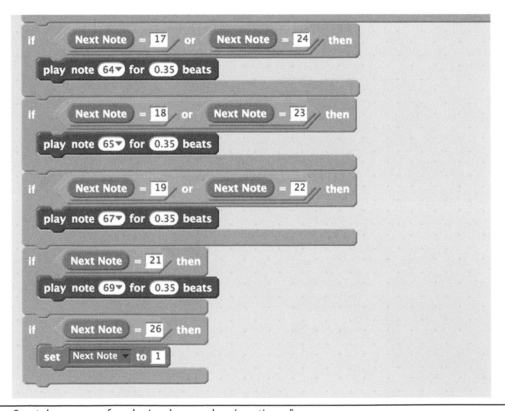

Figure 4-13 Scratch program for playing by number (*continued*).

right tones based on number. Also make sure to adjust the note duration in the "beats" section of the "Play note" block accordingly. We calibrated this program to produce each note with a short length to mimic the sound of a kalimba.

The last block you need for this program is to reset the counter. Drag one more "if/then" statement and a data block to "Set next note to 1" once the count equals 26. (If you write your own song, make sure to adjust the number here to match the number of notes you want your song to play.)

few long skinny leaves off of a houseplant. Place your leaves as in Figure 4-14, and place a piece of aluminum foil over the top. Use a thumbtack to hold each leaf to the box, but also ensure that all leaves will join the capacitance of the aluminum foil strip. Clip an alligator clip to a thumbtack inside the box, and get ready to play!

Taking It Further

What other items would make interesting one-button pianos? How could you combine this with other projects?

Create Kalimba

Step 1: Dig Up Supplies

Find a box that will fit comfortably in your hands. Dig up some backyard onions, or cut a

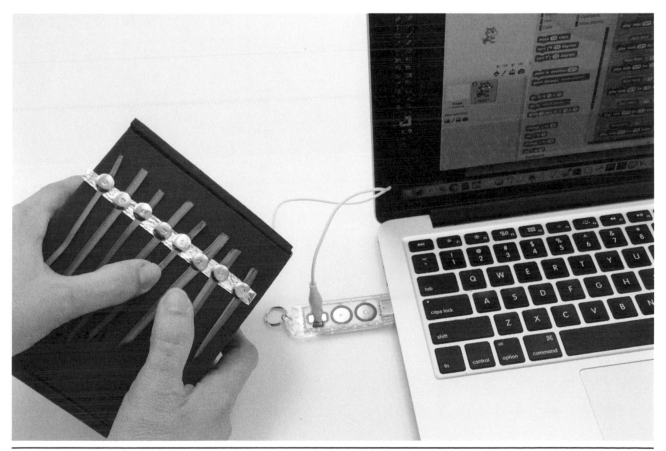

Figure 4-14 Kalimba go!

Makey Go Cat Clicking Game

Minions love click buttons. Why not turn something in your world into a giant button. We are going to use a giant conductive aluminum figurine we "borrowed" from our cat-loving neighbor. The player will try to pet the lion as many times as possible in 30 seconds. Players earn different cat levels from "kitten" to "lion" based on the number of times they pet the lion (see Figure 4-15). This game can be easily adapted to any type of conductive figurine to create a fun click-only game. Don't have a giant metal figurine lying around? You can easily cover a stuffed animal with aluminum foil and create your own.

Cost: $

Make time: 30 minutes

Skill level: 🍌🍌

Figure 4-15   Petting the lion.

Supplies

Materials	Description	Source
Conductive figurine	Metal figurine or any figurine covered with aluminum foil	Grandma's wall
Computer and access to Scratch	A free visual programming language and online community	scratch.mit.edu

Step 1: Backdrops and Sprites

In Scratch, click on the "Backdrops" menu. Create three backdrops, and use the "Fill with color" tool to paint each one a different color. The first two backdrops will just be used to trigger different scripts. The third backdrop will signal that the player's time is up. Select "Backdrop3," and use the "Rectangle" tool to create a large rectangle that is centered on the backdrop. With the "Text" tool, write the words "Time's up" over the rectangle. You can make the words larger by clicking and dragging the white squares that appear after you click outside the textbox.

With the backdrops complete, it's time to add a sprite. Click on the "Choose a new sprite" block from library icon located just below the "Preview" window. Choose the lion sprite from the animals' library, and then delete the default Scratch cat sprite. Click on the first lion sprite, and resize the lion so that it takes up the entire screen. Later we will use the "When sprite clicked" block so that we can make the sprite to take up as much room as possible. After you are done resizing, right-click, and duplicate the sprite. Click back on the first lion, put a large rectangle across the bottom of the sprite, and add the text, "Pet the lion once to start the game," as shown in Figure 4-16.

Figure 4-16   Lion backdrops ("Lion2").

Step 2: Lion Click Counters

Select the "lion" sprite, and then drag the script block "When this sprite is clicked" from the "Events" menu. Next, let's set up a variable to count the player's clicks. Select the "Data" menu, and then click on the "Make a variable" button. Name the variable "click#," make it available for all sprites, and then press "OK" (see Figure 4-17). Every time the game starts over, we want to set this variable to zero. To do this, attach the "Set variable to 0" to the "When sprite is clicked" block.

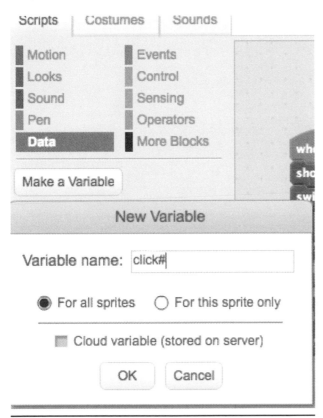

Figure 4-17 "Click#" variable ("Lion3").

Step 3: Instructions and Insurance

It's time to give our player some instructions. Attach the "Play sound until done" block to the existing script. Click on the dropdown arrow, and choose "Record." Give Scratch access to your microphone, and record the following statement: "When you hear the meow, start petting! You will have ten seconds to prove what kind of cat you are." Name this recording "Start petting." Click back over to the "Scripts" menu, and make sure that your new recording "Start petting" is selected. Add a "Wait 1 secs" block from the "Control" menu for a little suspense, and then add a "Switch backdrop block" from the "Looks" menu and select "Backdrop2." Follow this with a "Hide" block from the "Looks" menu. Switching backdrops will act as a visual cue for the player, while hiding the first lion makes room for the "Lion2" to count clicks. Add the "Play sound meow" block to the end of this block of code.

To ensure that everything resets while you are testing your code, add a "When green flag clicked" block from the "Events" menu, followed by a "Show" block from the "Looks" menu, as shown in Figure 4-18. Attach a "Switch backdrop to backdrop1" to the script. Drag a "When backdrop switches" block from the "Control" menu and a "Show" block from the "Looks" menu. This script ensures that the lion sprite reappears at the beginning of the game when it repeats.

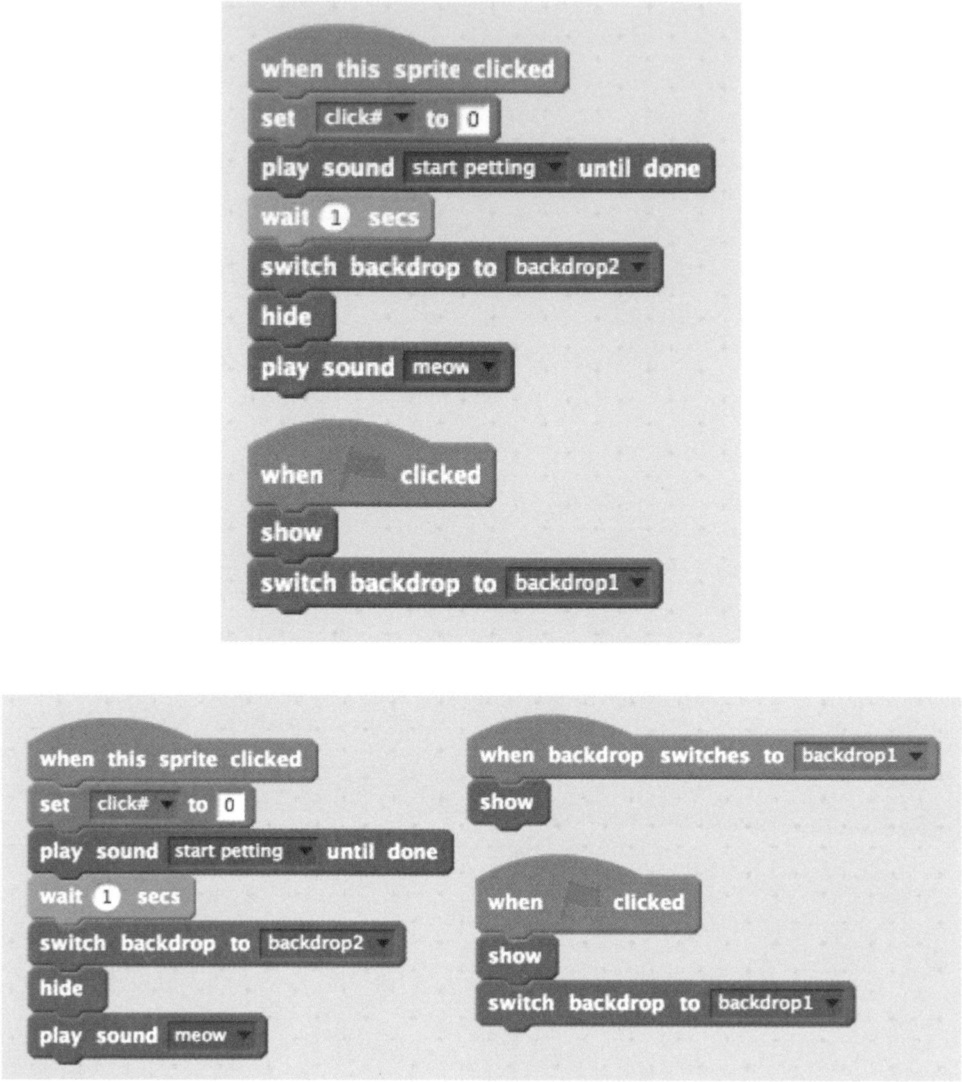

Figure 4-18 Start and reset scripts.

Step 4: Backdrop Switch and Click Count

Click on the sprite "Lion2," and drag a "When sprite is clicked" block to the scripts area. Add a "Change click# by 1" block to the script. This will add one count to the player's score when the "Lion2" script is clicked. To make the "Lion2" sprite appear when backdrop switches code block to the scripts area, select "Backdrop2" and add a "Show" block from the "Looks" menu to the script. To limit our player to 30 seconds, add a "Wait 1 secs" block to the script, and change the value to 10. Drag a "Hide" block to the

script next so that once 30 seconds are reached, the player can't click on it and change the score anymore. To provide a visual cue for the player, add a "Switch backdrop to" block and change the value to "Backdrop3."

Add three "if/then" blocks from the "Controls" menu to the script. Drag the operator "□ < □" into the condition value. Drag the variable "click#" into the first value, and input 20 for the second value. For the "then" part of the condition, add a "Play sound until done" block from the "Sounds" palette. Click the dropdown arrow in the "Play sound" block to choose "Record" so that you can record this statement, "That was pawsitively pawful! Cute kitten level fur you!" and name it "cute kitten."Now, during game play, if the player pets the lion fewer than 20 times (which will register as 20 clicks), this message will play.

For the next "if/then" block, drag the "and" operator into the statement. In the first value, place the "□ > □" block. For the first variable, add the "click#" variable and for the second input the number "19." For the next spot in the "Condition" block, add a "less than" operator. Drag a "click#" variable into the first spot and input "35" into the second. Inside the "then" portion of the block, add a "Play sound until" block, and record another hilarious statement—something like, "Wow, that was fast. You must be some kind of cheetah, with cattitude!" Name the recording "cheetah." If a player's score

falls between 20 and 34 points, this statement will play. For the last "if/then" block, add a "greater than" operator ("□ > □"), and input the variable "click#" into the first spot and the value "34" into the second (see Figure 4-19). Add a "Play sound until done" block inside the "then" portion of the condition. Then record the last punny statement, "Perfect score! You're a lion no doubt, king of the jungle." and name it "lion." To give the statement time to play, add a "Wait 1 secs" block and change the value to "3." You may need to adjust this time depending on your recording length. Follow this block up with a "Switch backdrop" block from the "Looks" menu, and select "Backdrop1." For testing and restart purposes, add a "When green flag clicked" block followed by a "Hide" block and then a "Switch backdrop to backdrop1" block.

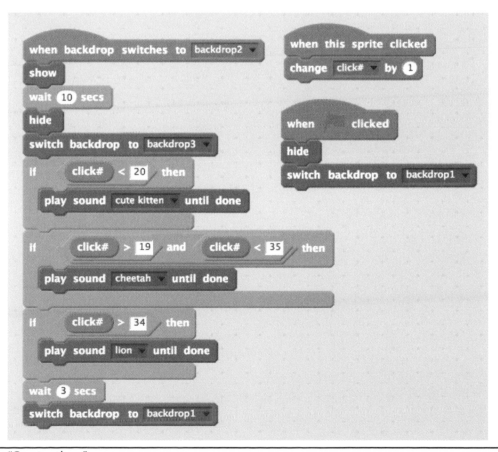

Figure 4-19 "Greater than."

Figure 4-20 Petting with "cattitude."

Step 5: Plug and Play!

Now that you've got your one-button game created, it's time to hook an alligator clip from Makey Go to your figurine and start playing your game! Challenge your minions to a cat petting game (see Figure 4-20)!

Taking It Further

Think about what other actions you could trigger with "if/then" statements and counting. How could you create games to allow multiple players to participate and compare scores when the game finishes?

Makey Go Lemon Squeezy

EVIL GENIUSES WOULDN'T BE EVIL if they played fair, now would they? In this two-player Makey Go game there is no chance to win if you play it safe. The goal of "Lemon Squeezy" is for a player to squeeze a lemonade bottle as many times as possible in 30 seconds. However, either player can reset the score to zero at any time in the competition by squeezing a lemon (see Figure 4-21). Strategy and squeeze ability are sure to produce many sour looks!

Cost: $

Make time: 30 minutes

Skill level: 🍌 🍌 🍌

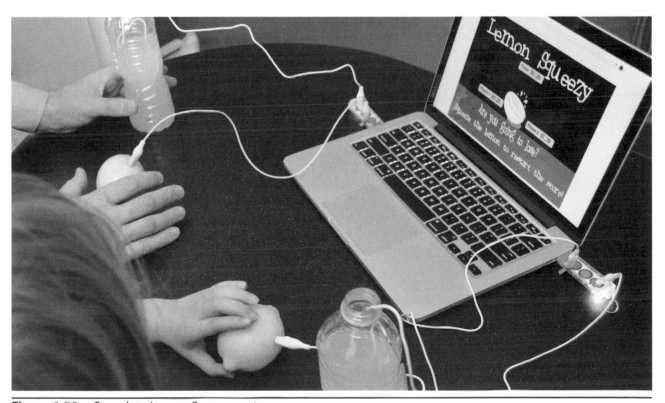

Figure 4-21   Complete Lemon Squeezy setup.

Supplies

Materials	Description	Source
Two Makey Gos	You will need two Makey Gos for this two-player game	Joylabz
Two plastic or glass bottles	Two plastic lemonade bottles or squeezable plastic water bottles	Grocery store
One lemon	Sour yellow citrus fruit	Grocery store
Computer and access to Scratch	A free visual programming language and online community	scratch.mit.edu

Step 1: Lemon Drops and Sprites

Create a new game in Scratch. Start by clicking the "Choose sprite from library" icon and adding the "Orange2" sprite. Click on the "Costumes" tab, and then use the "Color shape" tool to replace the color orange with yellow so that your orange can magically become a lemon! Click on the "Other costume" option to replace the color orange with yellow for all the other costumes (see Figure 4-22). After you finish, click on the "Information" tab for the "Orange2" sprite, and change the name to "Lemon." Position the "Lemon" sprite just

below the centerline on the stage. Right-click, and delete the "Cat" sprite. You will also need to add the following sprites from the letters library: "1-Pixel," "2-Pixel," "3-Pixel," "G-Pixel," and "O-Pixel" (see Figure 4-23).

To create a backdrop for instructions, click on the "Backdrop1" button in the lower left of the stage area. Click the "Convert to vector mode" button, and use the text and rectangle tool to create the backdrop shown in Figure 4-24. This backdrop will provide the players with directions for playing the game. Adjust the position of the lemon shape so that it appears in the gap

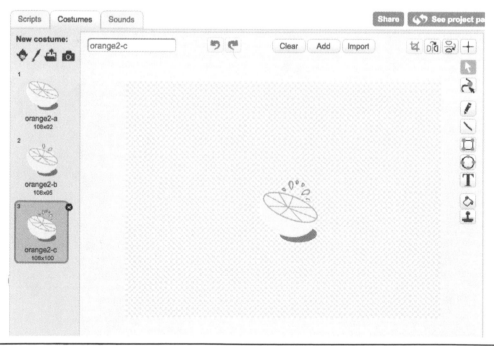

Figure 4-22   Turning an orange into a lemon.

Figure 4-23   List of sprites.

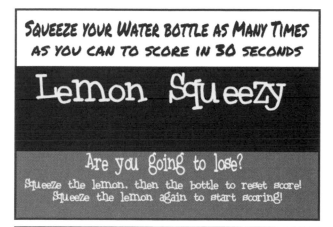

Figure 4-24   Instruction "Backdrop1."

between the text in the "Preview" menu. Right-click on "Backdrop1" to create "Backdrop2." Delete the instructions at the top, and replace them with a black rectangle, as shown in Figure 4-25.

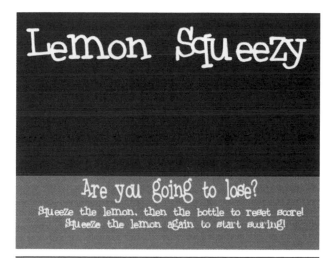

Figure 4-25   "Backdrop2."

You need to create two backdrops that will indicate which player won the game. Click on the "Paint new backdrop" icon to create a new backdrop like the one shown in Figure 4-26, and name it "Winner player 1." Right-click and duplicate the backdrop "Winner Player 1," and then change the name and text in the backdrop to "Winner Player 2."

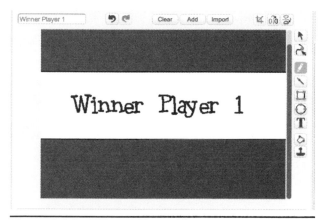

Figure 4-26   "Winner Player 1."

Step 2: Set Up Variables and Start Screen

Press on the "Lemon" sprite, and then select the "Scripts" tab. Click the "Data" menu, and click the "Make a variable" button to create the following variables: "Clicks," "Clickspeed," "Player1," "Player2," and "Timer." Drag a "When green flag clicked" block from the "Events" palette. So that the variables don't appear on the start screen, drag over three "Hide variable" blocks, and use the down arrow to set them to "Timer," "Player1," and "Player2." Click the "Looks" menu, and add a "Change backdrop" block to the script. Use the down arrow to select "Backdrop1." To add some animation and draw attention to the game, we can use the costumes for the "Lemon" sprite. Add a "Repeat 10" block from the "Control" palette, and also place a "Wait 1 secs" block inside the loop. Adjust the time in the "Wait" block to 0.5 second, and then add a "Next costume" block from the "Looks" palette.

Follow the loop up with a "Wait 1 secs" loop, and then reset the time to 5 seconds. To get the attention of your players, add a "Play sound" block, and choose a loud, abrupt sound from the sound library such as "Screech." The last block in the script will initiate the start of the game countdown by changing the background. Add a "Switch background" block from the "Looks" menu, and use the down arrow to choose "Backdrop2."

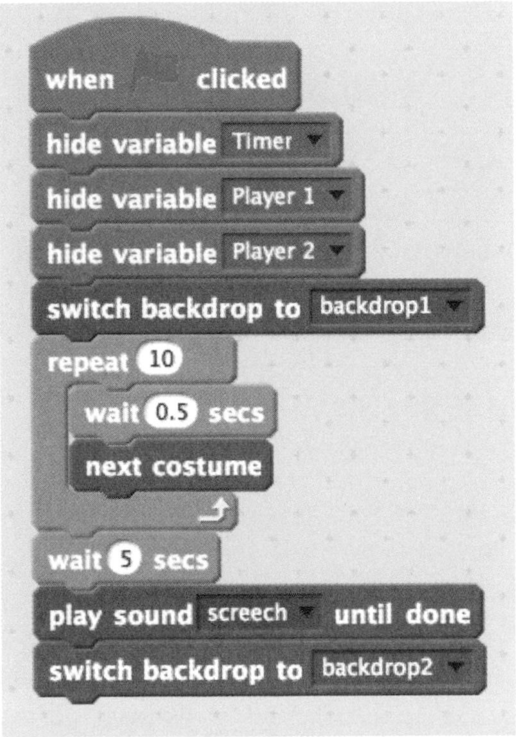

Figure 4-27 Start screen script.

Step 3: Variable Setting and Scoring Equations

With the lemon sprite selected, drag the "When backdrop switches to" block into the scripts area, and select the value "Backdrop2." Add a "Wait 1 secs" block to the script, and change the value to 4 seconds. From the "Data" palette, drag three "Set variable to" blocks to the script. In the first of those blocks, select "Player1," and set the variable to 0. Set the next one for "Player2" to 0. For the last variable, select "Timer," and set it to 30. At this point we want all three of these variables to show so that the players can keep score and know how much time remains. Add three "Show variable" blocks to the end of the script, and set them to "Player1," "Player2," and "Timer." To create a 30-second countdown, add a "Repeat 10" block from the "Control" palette, and set the value to 30. Drag a "Wait 1 secs" block into the "Repeat" loop, and then add a "Change variable by 1" block just before it in the "Repeat" block. Set the variable to "Timer," and enter "−1" into the second variable spot. Your countdown timer is now complete!

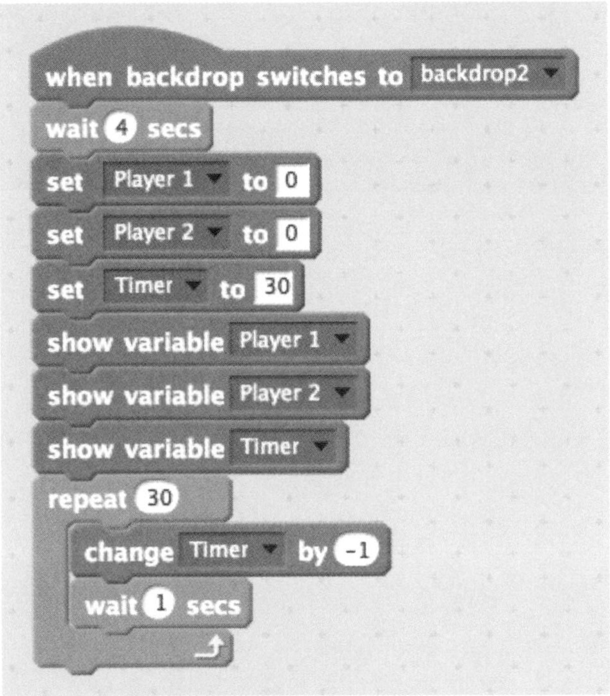

Figure 4-28 Variables and timer.

Now it's time to make some "if/then" statements that compare who has the most points and declare a winner by changing the backdrop. Drag an "if/then" block from the "Control" palette, but don't attach it to the script you have been working on yet. Leaving it disconnected will allow us to easily duplicate it and change the values. Drag the operator block "□ = □" into the "if/then" statement. We need to tell the script to perform these comparisons when the value of the timer is 0. So drag the "Timer" variable into the first part of the operator, and input the value 0 in the second spot. To start our comparison, drag an "if/then" statement into the first "if/then" block. Place the operator block "□ > □" into the "if/then" block. To compare the players' scores, drag the variable "Player1" in the first location and "Player2" in the second. If Player 1's score is greater, we need to declare him the winner. To accomplish this, drag a "Switch backdrop" block from the "Looks" palette inside the "if/then" statement, and change the value to the backdrop "Player 1 Winner." To add some animation, add a "Repeat 10" block just after the backdrop switch. Place a "Next costume" block inside the "Repeat 10" block followed by a "Wait .5 secs" block.

Now Player 1 can win the game, but we also need to provide the same conditional statement and argument for Player 2. Right-click on the section of code you just created for Player 1, and click "Duplicate." Switch the order of Player 1 and Player 2 in the "if/then" operator to "If Player 2 > Player 1." The last thing you will need to do is switch the backdrop to "Winner Player 2." With these changes in place, attach the two script blocks to the script that begins with the switch to "backdrop2" (see Figure 4-29).

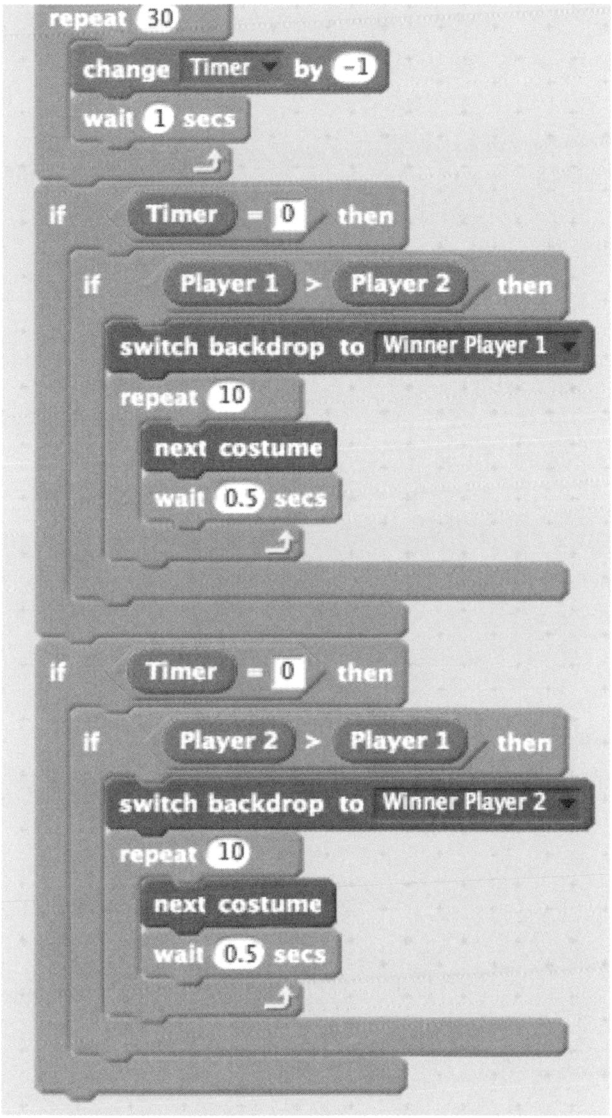

Step 4: Countdown Clock

To create a "3, 2, 1, GO countdown," we are going to use the "Show" and "Hide" blocks from the "Looks" palette as well as the "Wait 1 secs" block to time each sprite. Let's start with the "3-Pixel" sprite. If the sprite is not showing, click on the "i" to bring up information about the sprite. Click the box next to the word "show," and it should appear. If you want to only work with this sprite, you can click on the "i" for each of the other sprites and unclick the box. Click on the sprite that appears on the stage, and drag

Figure 4-29 Are you a winner script?

it to the position $x = 0$ and $y = 30$. You will need to do this for each of the number sprites in the countdown so that they appear centered just below the timer. To get the "G" and "O" into the correct position, place the "G" at $x = -26$ and $y = 30$ and the "O" at $x = 9$ and $y = 30$.

For testing purposes, you need to drag over a "When green flag clicked" block to the scripts area and follow it with a "Hide" block from the "Looks" menu for all the number sprites and the letters "G" and "O." An easy way to copy it over to the other sprites is to place it in the backpack and then select the sprite and drag it onto the scripts area. Once you have added this script, click on the "3 - Pixel" sprite. Drag over the "Event" block "When backdrop switches to," and change the value to "Backdrop2." Follow this block with a "Show" block and then a "Wait 1 secs" block. Because this is the first sprite to appear in the countdown and we only want this sprite to appear for a second, place a "Hide" block at the end of the script, as shown in Figure 4-30. Click the up arrow on the backpack, and drag this script into the backpack because we will need most of it for the other countdown sprites. Select the "2-Pixel" sprite next, and drag the previous script into the script area. We still want this sprite to appear for 1 second, but we need to delay when it shows up by adding a "Wait 1 secs" block just after the "Switch backdrop to backdrop2" block. Copy the script into the backpack, and then select the "1-Pixel" sprite. Move the script over, but this time change the "Wait 1 secs" block value to 2 seconds. For the "G-Pixel" and the "O-Pixel" sprites you will need to change the value of this block to 3.

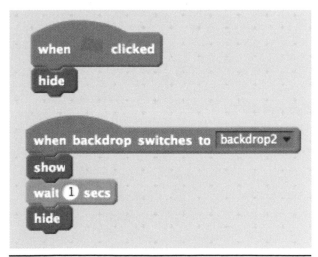

Figure 4-30   Script for "3-Pixel" countdown.

Step 5: Click and Clack

Just to make this game super evil, you are going to add a way to reset the scoring at any time during the game. For this, you'll have a lemon in between players that players can tap to switch the Makey Go to "Click" mode, tap again to "Click" the backdrop to reset the score, and then tap the lemon one more time to start scoring again. To make this effective with our one-button constraint, we'll write this program in the "Stage" area on the "Backdrop." All you need to make this work is an "Event" block "When stage clicked" with two data blocks "Set player1 to 0" and "Set player2 to 0," as shown in Figure 4-31. You'll have to attach some conductive tape to the gear input on the Makey Go. In this way, either player can tap the lemon and set one of the Makey Gos to "Click" mode, and then the player will have to squeeze the water bottle to set the game back to 0. Then tap the lemon again to put it back into scoring mode. (It will be important when playing that you move your mouse from the green flag to the center of the stage! Otherwise, you'll just be clicking the game to start over and over and over.)

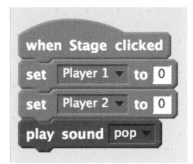

Figure 4-31   Reset scripts.

Step 6: Connect and Play Tips

With all the code complete, it's time to get two Makey Gos going! Cut a 1½-inch length of conductive tape, and fold about ½ inch of it over on itself to create a tab. Press the rest of the tape

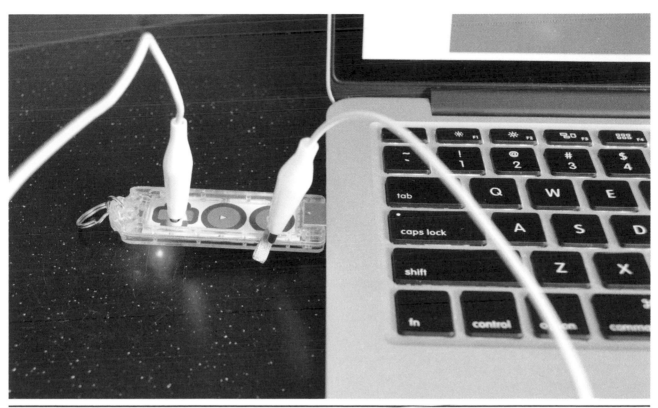

Figure 4-32   Makey Go setting hack.

across the "Gear" button on your Makey Go, as shown in Figure 4-32. Connect an alligator clip to the tab, and push the other end into a lemon. Repeat this step for the other Makey Go. Plug both Makey Gos in simultaneously. Choose one, and hold the lemon for 5 seconds to turn on remapping. Remap space to the "a" key. Test both by tapping the plus sign to see if the game is keeping score. It's important for all players to know that if they hold the lemon instead of squeezing and releasing it, they may accidentally cause it to go into remap mode. It's also important to plug into your lemon before you plug the Go into your computer.

To complete the game, grab some lemonade bottles and take a swig or two; then drop an alligator clip in them. Connect the other end to the "Plus" key on Makey Go, and press the "play" button. Now you are ready to squeeze.

Taking It Further

Start thinking of other ways you could use two Makey Gos at once! Is there a way to make this same game using the classic Makey Makey? What are fun ways you could combine the need for two capacitive controllers?

Index